Plant Microbiology

R. Campbell

Lecturer in Microbiology,
Department of Botany,
University of Bristol

Edward Arnold

First published in Great Britain 1985
by Edward Arnold (Publishers) Ltd
41 Bedford Square, London WC1B 3DQ

Edward Arnold (Australia) Pty Ltd
80 Waverley Road
Caulfield East
Victoria 3145
Australia

Edward Arnold,
300 North Charles Street
Baltimore
Maryland 21201
U.S.A.

British Library Cataloguing in Publication Data

Campbell, R.
Plant microbiology.
1. Micro-organisms 2. Micro-organisms,
Phytopathogenic
I. Title
576 QR41.2

ISBN 0-7131-2892-5

Text set in 10/11pt Plantin
by The Castlefield Press
Printed and bound in Great Britain by
Thomson Litho Ltd, East Kilbride, Scotland

Preface

This book is about those micro-organisms that are associated with plants, especially with vascular plants in terrestrial environments. It is inevitably biased towards agricultural crops of temperate regions of the world which are the only plants whose microbiology has been extensively investigated. An attempt has been made to include saprotrophs and biotrophs, pathogens and harmless or beneficial organisms in one discussion. It is not therefore intended that a plant pathologist will be able to study details of this or that disease, but rather that students of botany, zoology, microbiology, agriculture and horticulture will gain an overall impression of the micro-organisms on, in or near plants.

The first two chapters introduce micro-organisms to those unfamiliar with them, but no microbe is described in any great detail and by no means all organisms associated with plants receive a mention. Gilbert White (*A Natural History of Selborne*) said in 1788 that 'the standing objection to botany has always been that it is a pursuit that amuses the fancy and exercises the memory, without improving the mind or advancing any real knowledge: and where the science is carried no farther than a mere classification the charge is but too true'. The same might be said today of some branches of microbiology, so long lists of species have been avoided as much as possible. However the opposite extreme must also be avoided. Microbiology is not purely the province of biochemists and molecular biologists with microbes looked at as just convenient sources of enzymes and genes; micro-organisms live in the world outside the laboratory. I have tried to write a natural history of the fungi, bacteria, protozoa, algae and viruses associated with plants; unfortunately this aim is defeated on some occasions, because the information available reflects what commercial and research organizations are interested in or what is a fashionable field of study, rather than what is really happening in natural environments.

The text does not contain extensive references, and I apologize to those who recognize their work which has not been specifically acknowledged. However, references are given to a few outstanding papers in each subject and each chapter has a list of references designed to lead the reader into the more detailed literature. Tables and figures are fully referenced and have on occasion been chosen to highlight the work of the authors, as well as to illustrate the matter under discussion. I hope that, despite some errors which must have crept in, the readers will find much to interest them and that above all they will enjoy reading this book, even if it is on their required reading list! I am not especially concerned that large numbers of facts will be committed to memory, though this may be a laudable aim; I should much prefer that the diversity and importance of micro-organisms in the world arouses the reader's interest and enthusiasm.

I would like to thank all my colleagues in the Departments of Botany and Zoology, University of Bristol who have withstood my tedious enquiries, and who have given of their knowledge, their opinions, their photographs and perhaps most important, their time. In particular Dr Alan Feest read the manuscript and offered much helpful advice, Dr A. Beckett read the proofs, Jean Hancock typed a large part of the text and Tim Colborn helped with photography. Sue Pettit and James A.H. Brooks succeeded in satisfying my considerable demands on the library facilities, finding books which my inadequate referencing system, and even worse memory, had made almost unrecognizable.

Bristol, 1985 R. Campbell

Contents

1
Introduction

The limits of microbiology are difficult to define: it is the study of micro-organisms but the problem is to decide which organisms should be included. Obviously they should be small, usually needing at least a light microscope to see them properly. Viruses, bacteria (including mycoplasmas, rickettsia and actinomycetes), cyanobacteria (blue-green algae), fungi, protozoa and some small eukaryotic algae are usually considered to be included in the definition of micro-organisms.

This book is concerned with all micro-organisms that live in association with higher plants, usually terrestrial plants. As long as it is possible to classify them to some extent, so that a name can be given to them, it will be possible to make some generalizations about their characteristics and properties. There is however much work still to be done on classification and taxonomy, especially with the bacteria, and not all organisms associated with plants have a name. Even with these limitations a brief outline of the main taxa of organisms which will be the subject of later chapters is given. More details can be found in standard microbiology texts (Stanier, Adelberg & Ingraham, 1976; Brock, 1979) and in the standard texts on particular groups (Buchanan & Gibbons, 1974 for bacteria; Starr *et al.*, 1981 for prokaryotes; Ainsworth & Sussman, 1965–73 for fungi; Round, 1973 for algae; Sleigh, 1973 for protozoa).

There are major, and almost universally agreed, distinctions between viruses, prokaryotic and eukaryotic micro-organisms. Viruses are not cellular and usually consist of only nucleic acid and a few proteins: they are incapable of independent metabolism and are all obligate pathogens of prokaryotes and eukaryotes. A prokaryote is an organism lacking a nucleus and mainly having its DNA as a single, often circular molecule; examples are bacteria and cyanobacteria. Prokaryotes are always small, usually single-celled or at most filaments or groups of similar cells. Eukaryotes possess a nucleus with a nuclear membrane and chromosomes: usually there are also endoplasmic reticulum, mitochondria, Golgi bodies and, in the case of plants, chloroplasts. Protozoa, fungi, algae and all the higher plants and animals are eukaryotes. The main divisions of micro-organisms important to higher plants are summarized in Table 1.1. The microbes of greatest importance in plant microbiology are viruses, bacteria, protozoa and fungi.

Table 1.1 The main groups of micro-organisms that are important to higher plants.

Viruses No cellular structure or independent metabolism	Viruses which attack higher plants	Usually contain RNA, obligate pathogens of higher plants, algae and fungi. Classified on the basis of host and structure. Usually transmitted by invertebrates. See Table 1.2.
	Viruses which attack prokaryotes, called bacteriophage	Usually contain DNA, obligate pathogens of bacteria and cyano-bacteria, usually with a complex shape consisting of a head, tail and attachment mechanism.
Prokaryotes No nuclear membranes, a single, usually circular strand of DNA, no mitochondria	Bacteria	Mycoplasmas. No cell wall. Pathogens of plants and animals.
		True bacteria with walls (usually, but includes some strains or stages in life cycle that are without walls, e.g. *Rhizobium* inside roots). Organotrophs or photo- or chemo-lithotrophs; photosynthesis usually anoxy-genic. Single-celled or groups or filaments of similar cells. Classi-fied on wall chemistry nutritional characteristics and DNA chemistry. May be motile. Size range 0.2 μm to 100 μm; usually 0.5 μm to 3.0 μm See Table 1.3.
	Cyanobacteria	Oxygenic photo-autotrophs with chlorophyll *a* (can also be anoxygenic in some circum-stances). May be single-celled, collections of similar cells and some with 2 or 3 different cell types. Classified on morphology and type of cell division, also recently on DNA chemistry. May have gliding motility. Size 0.5 or 1.0 μm to 5.0 μm diameter by up to 50 μm or longer filaments.
Eukaryotes Cellular, with nuclei, mitochondria and other organelles. True chromosomes	Fungi	Organotrophs. Single-celled (yeasts) or filamentous. May produce macroscopic fruit bodies (e.g. toadstools). Classified on the basis of sexual reproductive structures. Usually non-motile but may have motile zoospores at some stages in their life cycle. Size of individual cells 5 μm to 10 μm diameter, 1 μm to 5 μm by several mm or cm long filaments, macroscopic hyphal aggregates usually a few cm but up to 1 m. See Table 1.4.

Table 1.1 *cont.*

	Protozoa	Organotrophs, often with complex requirements. Single-celled or acellular. Often motile by cilia or flagella. Classified on organs of locomotion, method of feeding and general morphology. Size 2 μm to 200 μm long.
	Algae	Oxygenic photo-autotrophs with chlorophyll *a* and *b*. Single-celled or filamentous or groups of similar cells to large complex multicelled organisms. Classified on pigment chemistry, general morphology, cell wall chemistry and sexual reproduction. Many single-celled algae are motile by flagella or gliding motility. Size 1 μm diameter to many metres long.

The main groups of micro-organisms

Viruses

Plant viruses cause a wide variety of diseases (Table 1.2) whose symptoms may be chlorosis of leaves in various patterns, stunting and distortion of the growth of the plant or parts of it. Plant viruses are often dependent on animals, especially arthropods such as aphids, bugs, leaf hoppers, etc., for transmission from plant to plant. It may be easier to control a virus disease by controlling the vector rather than the virus itself. In addition man now spreads viruses by grafting and mechanical damage in agricultural and horticultural operations and some viruses are systemic in the plant and are propagated in cuttings and seeds. The classification of the viruses is in some confusion. There have in the past been attempts, now abandoned, to use the Latin binomial system. The present system arranges viruses in groups (sometimes given the status of family) and defines that group in a cryptogram which codes for the types of nucleic acid and its molecular weight, the shape of the particle and its hosts and transmission mechanism. The main virus groups affecting higher plants are given in Table 1.2 (see also Gibbs & Harrison, 1976) and there are at least another 15 groups that are not included because they are mainly or wholly pathogenic to animals. Most of the plant viruses have RNA.

There are also some eight groups or families of viruses that are parasitic on bacteria. They contain DNA and are spherical or of very complex shapes with a head portion containing the nucleic acid in a protein coat and a tail protein, often with complicated attachment mechanisms. This complex type of bacterial virus is called a bacteriophage. Bacterial viruses are transmitted by

Table 1.2 The main virus groups of importance to plants.

Virus group	Shape	Nucleic acid	Vector	Examples of disease caused and group name derivation (by italics of key syllables)
Reovirus	Spherical	RNA	Aphids	Rice dwarf. Also multiplies in vector. (*R*espiratory *e*nteric *o*rphan virus)
Rhabdovirus	Elongate	RNA	Aphids, Leaf hoppers	Potato yellows dwarf. Also multiplies in vector. (Greek *rhabdo* = rod)
Caulimovirus	Spherical	DNA	Aphids	*Cauli*flower *mo*siac
Bromovirus	Spherical	RNA	Beetles	*Brome mo*saic, broad bean mottle
Comovirus	Spherical	RNA	Seedborne, beetles	*Co*wpea *mo*saic
Cucumovirus	Spherical	RNA	Seedborne, aphids, dodder	*Cucu*mber *mo*saic, peanut stunt
Ilarvirus	Spherical	RNA	Contacts, grafts	Tobacco streak, many others especially leaf discolourations of trees. (*I*sometric,*la*bile, *r*ingspot)
Luteovirus	Spherical	RNA	Aphids	Barley yellows dwarf, beet western yellows, soybean dwarf. (Latin *Luteus* = yellow)
Nepovirus	Spherical	RNA	Seedborne, nematodes	Tobacco ringspot, arabis mosaic, tomato ringspot, many others (*ne*matode transmitted, *po*lyhedral shape)
Pea enation mosaic	Spherical	RNA	Aphids	*Pea enation mosaic*
Tobacco necrosis	Spherical	RNA	Fungus	*Tobacco necrosis*
Tomato spotted wilt	Spherical envelope	RNA	Thrips	*Tomato spotted wilt*
Tombusvirus	Spherical	RNA	Not known or sap inoculation	*Tom*ato *bus*hy stunt
Tymovirus	Spherical	RNA	Beetle	*T*urnip *y*ellow *mo*saic
Alfalfa mosaic	Elongated	RNA	Seedborne, aphids	*Alfalfa mosaic* virus
Carlavirus	Elongated	RNA	Aphid (?)	*Ca*rnation *la*tent virus, potato virus M., cowpea mild mottle, many others
Closterovirus	Elongated	RNA	Aphid	Beet yellows, wheat yellow leaf, beet yellow stunt (Greek *kloster* = spindle or thread)
Hordeivirus	Elongated	RNA	Seedborne or unknown	Barley stripe mosaic. (Latin *Hordeum* = barley)
Potexvirus	Elongated	RNA	Mechanical contact or damage	*Pot*ato *X* virus, narcissus mosaic, cassava mosaic
Potyvirus	Elongated	RNA	Aphid or seedborne	*Pot*ato *Y* virus, beet mosaic, lettuce mosaic, plum pox, turnip mosaic, many others
Tobamovirus	Elongated	RNA	Mechanical contact, grafts, fungus	*Toba*cco *mo*saic, cucumber green mottle, tomato mosaic. Algae also a host
Tobravirus	Elongated	RNA	Nematodes, Dodder	*Tob*acco *ra*ttle

contact or are passed on from one generation to the next during division of the bacterium. They concern us mainly in that they may affect the ecology of agriculturally-important bacteria. Finally there are viruses which are parasitic in cyanobacteria, fungi and algae but these are incompletely described and have not been systematically classified. It is possible that they may eventually be located in existing families: for example some fungal viruses are close to the herpes simplex group of animal viruses and an algal virus is mentioned in Table 1.2.

There are agents of plant disease that were once thought to be viruses on the basis of their passage through fine filters. Some of these diseases have been shown to be caused by mycoplasmas and spiroplasmas which, though classified as bacteria, can pass through filters with a pore size of only 0.2 μm since they have no rigid walls. There are still many diseases assumed to be caused by viruses for which there is in fact no rigorous proof of the causal organisms.

Bacteria

Mycoplasmas, or mycoplasma-like-organisms (MLOs), are obligate biotrophs which cause several important plant diseases such as aster yellows and citrus stubborn. They are usually recognized only when diseased plant material is examined in the transmission electron microscope where they appear (in section) within plant cells as 0.1 to 1.0 μm diameter 'cells' bounded only by a unit membrane (see Gibbs & Harrison, 1976; Whitcomb & Tully, 1979).

The major groups of bacteria which are associated with plants (Table 1.1) are considered in more detail in Table 1.3. It should be stressed that only those groups of importance to plants are mentioned in this table, so many of the genera discussed in general microbiology texts, which are usually devoted almost exclusively to medical bacteriology, do not appear. The various attempts to classify the bacteria in traditional orders and families have now been abandoned by most bacteriologists. The different genera are merely grouped, either functionally as in 'the methane-producing bacteria' or on the basis of straining or biochemical tests used in classification (e.g. Gram-negative cocci). Such tests used in classification may or may not have importance in taxonomy and phylogeny. The main reference for descriptions and classification is *Bergey's Manual of Determinative Bacteriology* (Buchanan & Gibbons, 1974) and a more complete treatment is given by Starr *et al.* (1981), but some general texts have discussions of the subject (the best for bacteria from natural environments are Brock, 1979; and Stanier, Adelberg & Ingraham, 1976).

Gram-negative aerobic rods are particularly common on or near plants (Table 1.3) and the endospore-forming bacteria may be important in withstanding unfavourable conditions in the environment. Another common group are the pleomorphic, Gram-variable bacteria, the coryneforms, and the actinomycetes. The latter again form resistant spores.

The number of bacterial diseases of plants is really quite small, though some are important to particular crops. Most bacteria are either harmless saprotrophs (see p. 21) that live on the plant or decay organic matter and recycle nutrients. There are some special cases which we will consider where the bacterial breakdown of plant remains is utilized in agriculture or is of

Table 1.3 The main groups of bacteria of importance to plants.

Group (as in Bergey's manual)	**Shape and size**	**Examples of diseases, genera, species or importance to plants**
Mycoplasma-like organisms	Pleomorphic, no cell walls 0.1–1.0 μm diameter	*Spiroplasma citri* – citrus stubborn
Gliding bacteria	Rods and filaments or spindle shaped, 2.0–3.0 μm wide	*Cytophaga* – important in cellulose decomposition, and in lysis and degradation of other microbes
Spiral or curved bacteria	Spiral, 1–2 μm diam.,	*Azospirillum* – chemo-organotrophs, fix N in rhizosphere (p. 124)
	flagellate.	*Spirillum* – heterotroph, anaerobe, decomposition of carbon
	Curved, 0.2–0.4 μm diam.	*Bdellovibrio* – parasitic on other bacteria
Gram negative, aerobic rods and cocci.		
1. Pseudomonadaceae	Rods, polar flagellate, 0.5–1 μm × 1.5–4 μm	*Pseudomonas, Xanthomonas* – plant pathogens and very common leaf surface and soil saprotrophs
2. Azotobacteriaceae	Pleomorphic short rods or cocci. 2 μm diam.	*Azotobacter, Beijerinckia* – non-symbiotic N fixation
3. Rhizobiaceae	Rods, up to 3 μm long	*Rhizobium* – symbiotic N fixation in legumes *Agrobacterium* – crown gall in many plants
Gram negative, facultative anaerobes	Rods, up to 4 μm long	*Klebsiella* – soil saprotrophs, may fix N, plant pathogens *Erwinia* – plant pathogens (fire blight), common saprotrophs *Flavobacterium* – soil and leaf saprotrophs
Gram negative, chemoautotrophs	Short rods, usually not motile. 0.5–2 μm	*Nitrobacter, Nitrosomonas* – oxidation of ammonium to nitrate *Thiobacillus* – oxidation of sulphur
Methane producing bacteria	Coccoid or short rods. 0.7 μm × 0.8–1.8 μm	*Methanobacterium* – important in the rumen (p. 169)
Gram positive cocci	Cocci, 0.5–3 μm diam. Usually about 1 μm diam.	*Micrococcus* – common saprotrophs on plants *Ruminococcus* – cellulose breakdown in the rumen (p. 169)
Gram positive, endospore forming	Rods, 1 μm × up to 10 μm	*Bacillus* – aerobic, very common in soil and on plants *Clostridium* – anaerobic, cellulose breakdown, N fixation
Gram positive, non-sporing	Rods, 0.8 × 2 μm usually in chains	*Lactobacillus* – important in silage (p. 175)
Actinomycetes, gram positive or variable		
1. Corynebacteria	Pleomorphic	*Corynebacterium* – saproptrophs and plant pathogens. Very common in soil *Arthrobacter* – common saprotrophs in soil
2. Actinomycetales	Filamentous, about 1 μm diam.	*Frankia* – N-fixation in alder roots *Nocardia, Streptomyces* – common in soil, plant saprotrophs and pathogens

importance to farm animals (pp. 171 – 180). The importance of nitrogen fixation by free-living bacteria is in some doubt but symbiotic associations, especially with legumes, are essential to some agricultural systems as well as to natural environments. Classification schemes for bacteria have mostly been designed for medically-important organisms and they may not work very well on other groups. There are now systems which depend on the sequence of nucleotides and these produce different groupings, such as separating off, from the majority of the bacteria, the Archebacteria (methanogenic, some halophitic and some acid-tolerent bacteria). In addition some of the bacteria that we will consider have just not been studied in enough detail for their taxonomy to be understood and for good diagnostic methods to be developed.

Finally in the prokaryotes are the cyanobacteria which are of importance in plant microbiology in that they fix nitrogen, either as free-living or symbiotic organisms, in lichens for example. The classification of cyanobacteria is based on whether they are single-celled or some simple form of colony, and if the latter whether there are specialized cells such as heterocysts. They have recently been transferred from the *Botanical Code of Nomenclature* based on dead type specimens to the *Bacteriological Code* based on live cultures. Genera have been conserved in most cases but species are not easily distinguished in culture. Many of the ecologically-important types, which look quite distinct under field conditions and have been given species rank under the *Botanial Code* are rather difficult to separate in culture. Many of the species described and preserved as herbarium specimens do not have cultures. These problems will persist for some time, but eventually the new classification based on cultural characters and base ratios for their DNA (guanine-cytosine ratios) will put the cyanobacteria firmly with the other prokaryotes, the bacteria, instead of with the eukaryotic algae where they have previously been classified.

Fungi

The fungi are relatively easy to classify, at least to genus, for the classification system depends on the morphology visible with the naked eye or with the light microscope. The main groups are summarized in Table 1.4. Fungi are major plant pathogens, attacking all parts of the plant at all stages of growth and utilization, indeed plant pathologists have traditionally been mycologists and this perhaps partly explains why the classification and taxonomy of fungi associated with plants is so much better worked out than that for bacteria and viruses. There are also numerous saprotrophic fungi living on and in plants which may be merely harmless, or positively beneficial. The decomposition of plant remains (plant litter) is largely brought about by saprotrophic fungi and bacteria.

Algae and protozoa

The true algae (eukaryotes) are only of interest to microbiologists studying higher plants because they occur as epiphytes on leaves and bark. Their taxonomy is based on their morphology and on the photosynthetic pigments that they possess. The green algae have chlorophyll *a* and *b* and only two phyla concern us, the single-celled, flagellate Euglenophyta and the very large

Table 1.4 The main groups of fungi of importance to plants.

Group	Form	Occurrence
Ascomycotina Sexual spores borne in a sack-like ascus, usually in a conspicuous fruit body	Yeasts	Saprotrophic epiphytes, in soil, in flowers and nectaries (*Candida, Cryptococcus*). Fermentation of plant products to produce alcohol, bread, etc. (*Saccharomyces*)
	Filamentous	Many saprotrophs on leaves, in soil, on roots. Decomposition of organic matter and nutrient cycling. Many pathogens of roots stems, leaves and fruits (*Nectria, Erysiphe, Sclerotinia*)
Basidiomycotina Sexual spores borne on a basidium usually on a conspicuous fruit body	Yeasts	Saprotrophic epiphytes (*Sporobolomyces*)
	Filamentous	Many saprotrophs on roots and stems. Decomposition of wood (lignin and cellulose) especially. Many pathogens, especially on trees (*Heterobasidion, Serpula, Thanetophthora*)
Mastigomycotina Have motile zoospores		Mostly aquatic, but some occur in soil and a few are important plant parasites (*Phytophthora, Pythium, Peronospora*)
Zygomycotina Sexual spore a zygospore		Common in soil. Decomposition of organic matter, especially low molecular weight compounds (*Mucor, Rhizopus*)
Deuteromycotina No sexual stage known		Very common on or in all plant parts and in the soil. Saprotrophs important in decomposition and nutrient cycles. Many parasitic on roots, stems, leaves and fruits (*Cladosporium, Alternaria, Phoma, Penicillium*)

phylum, Chlorophyta (Round, 1973; 1981). The brown algae contain only one phylum which is frequently associated with plants, the diatoms (Bacillariophyta), which are found in soil and on moist plant surfaces.

The protozoa (Sleigh, 1973) contain two phyla of interest to us, the Sarcomastigophora which are flagellates and amoebae of many sorts which are commonly found in soil and the Ciliophora which are motile because of the rows of cilia (small hairs) on their surface. The ciliates are further subdivided on the basis of the arrangement of the cilia, the structure of the mouth region and the type of food which they use. Protozoa have been little studied on plants or in the soil. They are active under moist conditions and are important in grazing bacteria and increasing nutrient turnover rates. Specialized protozoa occur in some plant litter decomposition systems such as the rumen and the hind-gut of some termites (p. 167). Myxomycetes (slime moulds) and similar organisms are considered to be fungi by some microbiologists, but the active stages in the life cycle are amoebae or flagellate swarm cells classified by zoologists in the Mycetozoa.

The study of microbiology cuts across previously-accepted taxonomic boundaries, including that between obvious animals and plants, and ranges from the morphologically simple bacteria to the very diverse groups such as the fungi. There are, for example, algae and protozoa which are obviously related and are very similar morphologically except for the possession or lack of a chloroplast: given the right culture conditions it is even possible for some algae to lose or greatly reduce their chloroplast and become heterotrophic (see later). There have been attempts, other than those outlined above, to classify micro-organisms in such a way as to overcome some of these problems: for example there is a five-kingdom scheme (Margulis & Schwartz, 1982) which has (1) higher plants, (2) higher animals, (3) monera (the bacteria and cyanobacteria), (4) the fungi, and (5) the protists. Margulis and Schwartz have solved the problem of similar algae and protozoa by grouping them together in the protists but this does not completely solve the problem of diversity within a group: what for example is to be done with the algae which are in the protists, but range from microscopic, unicellular organisms (clearly micro-organisms), to very large seaweeds such as kelp and *Sargassum*. There seems no clear solution to some of these problems, even though it is confusing to those used to the comparatively clear cut pigeon-holes of higher plant and animal classification. For our purpose it does not really matter how the theoretical taxonomists seek to overcome these problems of relationships between organisms and how the groups have evolved from each other, i.e. their phylogeny.

Microbial nutrition and physiology

There are other ways of looking at the grouping of micro-organisms apart from the taxonomic ones just discussed, for example they may be grouped according to their nutritional characteristics. The simplest of these is the distinction between heterotrophs, requiring existing organic matter as an energy source, and autotrophs which utilize light or inorganic chemical energy to produce organic matter from inorganic compounds such as carbon dioxide and water. Those that use light energy are photo-autotrophs (such as algae and cyanobacteria), and those using inorganic chemical energy are chemo-autotrophs such as the sulphur oxidizing *Thiobacillus*.

The heterotrophs cover an enormous diversity of organisms including most bacteria, all the fungi and most protozoa. They may be broadly divided into saprotrophs which live on detritus which is already dead (they are also called saprophytes, though this should be applied only to organisms living on plant remains not to those living on dead organic matter in general). Necrotrophs invade and kill a plant and then live on the dead material. Thirdly, biotrophs live on or in living plants and may have very complex nutrient requirements. Even amongst the saprotrophs there is an enormous range from those which need only sugars, inorganic nitrogen and mineral salts (many bacteria and fungi) to those which have a requirement for several energy sources, a variety of amino acids, vitamins, sterols, etc. (most protozoa and some fungi and bacteria). There are those whose nutrition is so complex that they have not been grown in artificial culture and these are sometimes called obligate parasites (but see p. 21). Many microbes can only be grown on undefined media such as extracts of plants or animals like heart-brain infusion agar, blood agar, potato or oatmeal agar. All organisms need a wide range of chemicals for their metabolism; the differences between organisms in their nutrient requirements are a reflection of their ability to synthesize what they need. Those with simple requirements can synthesize all they need from inorganic molecules and those with complex requirements cannot synthesize amino acids, sterols or whatever and must obtain them ready-made.

Most saprotrophs associated with plants and other natural environments have quite simple requirements and grow best at low concentrations of nutrients. Many of the media routinely used in mycology and medical bacteriology are not suitable for use with isolates from plants since the very high levels of available carbon and nitrogen, in Nutrient Agar for example, may well inhibit or even prevent growth. This may be a toxic effect or an effect of a low osmotic potential: most microbes associated with plants grow in water films or in very dilute solutions.

This is a part of a more general property of micro-organisms from plants; the levels of temperature, pH, etc., which are discussed in many microbiology texts are biased towards medically-important bacteria or laboratory strains of fungi. The most obvious example of this is temperature: bacteriologists, especially, tend to grow all their isolates at 37°C and mycologists routinely use 25°C. These temperatures are suitable for isolates from humans or from laboratory or industrial fermenters but they are not suitable for plant isolates which, in temperate countries at least, generally grow at 10–15°C. There are of course organisms which tolerate or require higher temperatures up to 60 or even 90°C (thermo-tolerant or thermophilic) but these are never found associated with living higher plants, though they may occur in heaps of rotting plant litter. Similarly there are cold-tolerant or cold-requiring isolates (psychrophilic) but again they are uncommon either in or on plants and even in severe winters or in the polar regions most of the microbes are not specially adapted to cold, they merely become dormant in unfavourable conditions.

The pH of plant substrates and the soil in which they grow is usually quite close to neutral. There are extremes; in some peatland vegetation the soil pH may drop to 4 or even 3, and a few plants grow in alkaline deserts. In general a pH of 6–8 will be the best for the growth of most microbes associated with plants.

Light may be important, photo-autotrophs obviously need it, but for most of the heterotrophs there is a danger of exposure to ultra-violet (UV) in sunlight. Many leaf surface microbes are pigmented, probably as a protection against UV. Light may also be important to some fungi in controlling sporulation in various ways (see Hawker & Linton, 1979, for a discussion of this and other physiological factors; also Brock, 1979; Dawes & Sutherland, 1976; Stanier, Adelberg & Ingraham, 1977). Light also has a secondary effect since its intensity affects photosynthetic rate and hence the amount of fixed carbon that a plant may have available for saprotrophs or pathogens.

The micro-environment is certainly aerobic over most of the surfaces of leaves because of photosynthesis. Around roots the situation is different for there are many micro-habitats in the complex mixture of materials that make up the soil. Anaerobic conditions, or at least low oxygen tensions, can be created by the respiration of the roots or of the microbes decaying organic matter (Campbell, 1983). Wet soils are especially liable to have anoxic or micro-aerophilic conditions because of the slow rate of diffusion and low solubility of oxygen in the moisture films.

Methods in plant microbiology

All these physiological attributes control the methods used in plant microbiology. The general maxim is to follow the conditions of growth which are present where the plant host grows. For heterotrophs media low in organic carbon and nitrogen, probably based on plant extracts, are used and incubated at suitable, usually low, temperatures. Protozoa may need to live on bacteria or yeast colonies. There are many books which give detailed methods for isolating and handling micro-organisms from plants and their surroundings (Johnson & Curl, 1972; Aaronson, 1970; Primrose & Wardlaw, 1982; Burns & Slater, 1982; Gibbs & Harrison, 1976).

Viruses can of course only be grown in living plants and sometimes only in their particular host. The investigation of a suspected virus disease usually involves inoculating a range of laboratory-grown virus-free plants. The type of virus can then be identified by the symptoms produced on the known hosts. Examination by electron microscopy of negatively-stained virus suspensions or of sections of infected tissue is also used for determining the shape of the virus and its location within the plant. Serological techniques can confirm the identification by comparison with known antisera.

Other obligate plant pathogens may need to be handled in a similar way to viruses. Many of the rust fungi cannot be grown in culture and most must be maintained by inoculation of the appropriate host plant. Methods of isolating organisms other than obligate pathogens are very diverse. If the growth is extensive, as in the rotting of a fruit, then it may be possible to just pick off or cut out a small piece of the macroscopic colony and plate it out on a suitable medium. Such microbes may be saprotrophs or necrotrophs and it is essential to prove that the organism(s) isolated really cause the disease and are not just chance contaminants: the fulfilling of Koch's postulates proves that the right organism was isolated. To do this the pathogen is isolated, grown in pure culture and identified (Fig. 1.1). It should be the same as the organism present in the diseased tissue if this could be identified. The pure, identified isolate is then inoculated into a healthy plant and must produce the same symptoms as the original disease, and the same isolate must be obtained again from the artificially-inoculated diseased tissue.

For surface-growing organisms the plant part can be swabbed or washed vigorously by shaking in water and the resulting suspension can be diluted and plated out. This allows the separation of different components of a mixed flora and is commonly used for leaf, root and seed surface micro-organisms. It tends to isolate mostly bacteria and spore-bearing fungi. It can be made quantitative on an area or tissue-weight basis and the organisms may be identified from the cultures. This general dilution plating method can also be used on macerated plant tissue if it is not necessary to separate those organisms growing on the

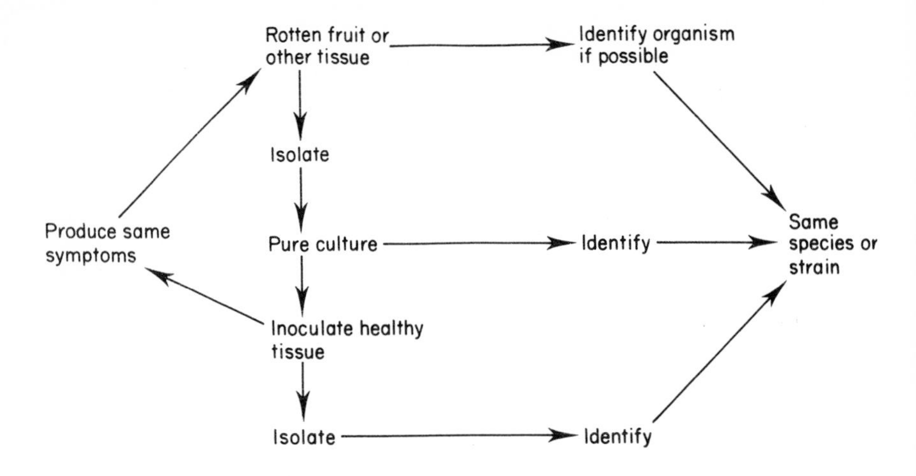

Fig. 1.1 The sequence of isolations needed to prove Koch's postulates, that is to show that a particular organism is responsible for a particular disease.

surface from those growing within the plant and if the microbes can withstand the maceration process.

If it necessary to isolate organisms growing inside the plant as pathogens or mutualistic symbionts (p. 14) without getting contamination by surface organisms, then the surface must be sterilized first. The seed, root or leaf can be soaked in a variety of solutions such as hypochlorite, silver nitrate or mercuric chloride and then the isolation made from the inside of the specimen. A balance has to be struck between the rate of penetration of the sterilant and the time required to kill the contaminants. If there is too much time all the organisms, including those growing inside the plant will be killed. If the time is too short then the culture will be contaminated. Subsequent cultures may need to have any remaining contaminants removed by plating on selective media such as those containing antibiotics known to inhibit the particular contaminants present, but not the required organism.

There are also cultural methods dependent on enrichment techniques which are used when the required organisms are infrequent or too slow growing or too specialized to be obtained with normal media. The procedure varies with different organisms but attempts are made to provide conditions which will favour the desired organisms (elective cultures) or prevent the growth of others (selective cultures). For example there are important bacteria that convert NH_4^+ to NO_3^- near plant roots. These organisms are obligate aerobes and chemo-autotrophs so the enrichment culture for them has NH_4^+ and some mineral salts and is incubated in the dark. There is no organic carbon source so heterotrophs will not grow, and no light so photo-autotrophs are eliminated. The solution is kept well aerated and inoculated with a little soil. The growth of the organisms can be monitored by the disappearance of NH_4^+ and when this occurs the solution may be plated out on the same medium solidified with agar and the numbers will now be great enough to give a good chance of isolation.

Apart from knowing how many and what kinds of micro-organisms are associated with plants, which can be done by cultural methods, it is often necessary to know where on the plant they were growing. Obviously we can tell quite easily whether it was growing on a root or a leaf, but finding out where on the leaf, root or seed may be a little more difficult and depends on direct observation with the light, or perhaps the scanning electron microscope. It is necessary to make small pieces of tissue transparent for the light microscope by clearing them in organic solvents or in chloral hydrate to remove the chlorophyll, before staining them to show the organisms (with phenyl acetic aniline blue for example). Alternatively surface illumination, possibly with the use of UV light and special fluorescent stains (epifluorescence), can be used for opaque material or larger structures. Such examination tells you where the organisms grow, and they can of course be counted.

The final possibility for the assessment of micro-organisms is to try to measure their metabolic activity. With micro-organisms associated with plants this can be difficult because of the confusion caused in the attempts to separate the plant's metabolism from that of the microbe. However it is possible, by the use of radioactive tracers, to measure microbial respiration rates. The microbes may also be physically separated from the plant in some instances and then studies made of their activity by assessing ATP or some particular enzyme. These methods which measure some metabolic process tell something about activity of micro-organisms but they do not allow them to be identified and counted, or give information on the micro-habitat.

In studying micro-organisms associated with plants it is important to consider very carefully just what data are needed before you design the experiment. The micro-habitats on and around plants are so varied, and change over such short distances, that it is sometimes very difficult to arrive at a suitable sampling plan. The work involved in detailed studies often requires rather tedious plating out or counting by direct observation of a great many different samples. It is important to notice what methods are used in obtaining data when assessing published papers, for the method will bias the result. It is very likely that different methods will yield different numbers or different species lists. However there are now sufficient general methods to cover all common problems in plant microbiology and provided that the limitations of the individual procedures are recognized, useful results can be produced. Many of the methods used are simple and do not involve expensive equipment. So little is known about many basic aspects of plant microbiology that there is still plenty of scope for investigations using no more than a reasonable light microscope and the production or purchase of simple, routine media.

Microbial communities and successions

Up to now the isolation or culture of individual species has been considered, but this is not very realistic for they rarely occur in pure culture in nature. Microbes occur as mixed populations in communities. A population of micro-organisms is a number of the same, or at least similar, species. When many different populations are present and are interacting we have a community. The communities and their interactions with the abiotic environment constitute an ecosystem. A niche is the sum of all the environmental and organism characteristics which determine the role of a population in relation to others in the community. A niche is not a place, the place is the micro-habitat. In plant microbiology we will be concerned with the interactions of individuals and populations in micro-habitats and with the diversity of niches.

A lot of work has been done in trying to classify the interactions between individuals (see Alexander, 1971). Firstly organisms may occur in different habitats, or in the same habitat but in the different niches, and therefore they do not have any effect on each other. They may for example have entirely different food sources and be adapted to different micro-habitats so that they do not compete for nutrients or for usable space. Such a complete lack of interaction is called *neutralism* and is probably quite rare, though many interactions are so slight or so devious as to be neutralism for practical purposes. If there is an interaction it may be beneficial or harmful to one or both of the organisms or populations (Table 1.5). *Antagonism* is a general term for harmful interactions and may be subdivided into:

1. *Competition* in which both organisms are harmed: for example by both getting less nutrients because both are exploiting a limited resource.
2. In *amensalism* one individual or population is harmed, by an antibiotic for example, while the other is not affected directly.
3. *Parasitism* and *predation* require that an individual benefits by harming another, usually by using the 'host' as food: this may involve damage or death of the host.

The beneficial interactions are no less varied. If only one organism benefits then the relationship is *commensalism* and the most usual example of this is the production of a vitamin or available nutrient source by one organism to the benefit of another. When both organisms benefit, the interaction is *mutualism* which is probably the most studied form of symbiosis. Examples which will be covered later include leaf and root nodules and a wide range of microbial systems which decay plants for an herbivorous host (e.g. termites, some ants, all mammalian herbivores).

The situation is, of course, not quite so simple as this classification suggests for there are many interactions where it is not at all clear who gets what from whom. This is often merely a reflection of our ignorance, but there may well be genuine cases where the type of relationship varies with time or is in a state of flux. There is some further confusion over the use of the term 'symbiosis'. As originally defined by de Bary it meant any two organisms living together, so it may include many of the categories of interaction listed above. This is the sense in which it should be used, but for many it has come to describe just mutualistic associations. It will be used in its original, broad sense in the chapters which follow and it may be qualified when the relationship is understood, e.g. mutualistic symbiosis. This has the advantage that when the benefits or otherwise of the relationship are unclear the description symbiosis does not require it to be specified.

Most of the symbioses which we will consider will be between microbes and higher plants but there are many others of ecological importance which are microbe–microbe interactions (see Campbell, 1983).

These interactions between individuals or populations may change with time. Thus there are different groups of micro-organisms on roots, leaves, etc.,

Table 1.5 Interactions between individuals or populations. For simplicity only two individuals or populations, A and B, are shown. Under natural conditions many would be interacting simultaneously with each other.

Individual or population interaction	Interaction name	Example
A → B: No effect; B → A: No effect	**Neutralism**	A and B in different habitats or at very low population levels
A → B: Benefits; B → A: No effect	**Commensalism**	A produces vitamin for B
A → B: Benefits; B → A: Benefits	**Synergism**	Not obligatory. Exchange of metabolites in green sulphur bacteria/sulphur reducer consortia. Complimentary enzyme systems in polymer degradation
A → B: Benefits; B → A: Benefits	**Mutualism**	Usually obligatory dependence of organisms. Bacterial N-fixing root nodules. Lichens
A → B: Harms; B → A: Harms	**Competition**	Both A and B using same limited food source
A → B: Harms; B → A: No effect	**Amensalism**	A produces antibiotic which harms B
A → B: Eats; B → A: Benefits	**Predation**	A (a protozoon) eats B (a bacterium). A usually larger than B
A → B: Harms; B → A: Benefits	**Parasitism**	A lives on or in B and uses B as a food source. A usually smaller than B. B may or may not be killed

which change as the plant grows, matures, senesces, and finally dies and decays: there is a succession of interacting populations with time. These changes in populations, and therefore in communities, do occur, though how and why is not always clear. The changes may be brought about by the activities of the organisms themselves: perhaps the most common example is the use of a nutrient or the production of toxic waste products so that the organism dies and is replaced by one that is better adapted to the new nutrient conditions or the different environmental factors, such as pH. Alternatively the succession may be a seasonal one, controlled by changing weather patterns outside the microbial community itself.

Table 1.6 A comparison of resource and seral successions. Two different sorts of resources are considered, one nutrient rich and the other nutrient poor. There are obviously exceptions to the generalizations made, but a framework is provided for the study of successions. (Based on personal communication (Swift, 1983) and Swift, M.J., 1982, in R.G. Burns, & J.H. Slater, *Experimental microbial ecology*. Blackwell Scientific Publications, Oxford. p. 164–77.)

	Ecosystem, seral succession		Resource, decomposer or substrate succession			
Ecosystem attributes	Rock or bare eroded soil →	soil	Resource with initially low available nutrients, e.g. wood, hard ever-green leaves, needing specialized enzymes →		Humus	← Resource with initially high available nutrients, e.g. soft fruit or animal body
Stage in succession	Early →	Late	Early →	Mid →	Late	← Early
Organic matter	Low →	High	High but unavailable polymers →	High and diverse breakdown products →	Lower, humic and fulvic acids, and un-available	← High, available, simple sub-stances
Nutrients available to heterotrophs	Low →	High	Low →	Higher in 2nd succession →	Low	← High
Nutrient conservation	Low →	High	High →	Lower →	High	← Low
Minerals	Inorganic free →	Organic immobilized	Organic immobilized →		Inorganic mineralized	← Organic immobilized

Cycle of nutrients	Open, outside input and sinks	→	Closed	Closed	→			Open	←	Closed
Special heterogeneity	Low	→	High	Low	→	Higher	→	Low	←	High
Species diversity	Low	→	High	Low	→	Higher	→	Low	←	High
Niche specification	?Broad	→	?Narrow	?Narrow	→	?Broader	→	?Narrow	←	?Broad
Biochemical diversity	Low	→	High	Low	→	Higher	→	Low	←	High
Stability	Low	→	High	High	→	Lower	→	High	←	Low
Growth form of colonizers	*r*-selected	→	*K*-selected	*K*-selected	→	*r*-selected	→	*K*-selected	←	*r*-selected
Life cycles	Simple	→	Complex	Complex long	→	Simple Short	→	Complex long	←	Simple Short
Typical microbes	Lichens, algae, cyano-bacteria		Soil and litter fungi and bacteria	Basidiomycetes, ascomycetes with cellulase and polyphenol oxidase. Bacteria may be primary colonizers		Basidiomycetes, ascomycetes with bacteria in secondary successions		Basidiomycetes, ascomycetes actinomycetes able to break down humus		Deuteromycetes, zygomycetes bacteria, 'sugar organisms'

Though successional changes are thought of in terms of time, there may be no uniformity in the response. At any one time there will be variation within a resource especially if it be large like some leaves or the stems of woody plants. Thus different parts of a decaying leaf will be at different stages in the temporal succession: perhaps one part of the leaf was killed by a pathogen and the succession may be far advanced in that particular spot before the rest of the leaf dies.

These successions are the results of the colonization of different resources in the environment and they are called resource, substrate or decomposer successions. Even for large resources like a tree trunk, decomposition to humus takes only a few years, though the products such as humic acids may be in the soil for a very long time indeed (thousands of years). It is generally true that the rate of decomposition slows as the succession proceeds and the resource quality changes. Less easily decomposed materials accumulate at the end of the succession, finally leaving humus. Again this decrease in availability may not be uniform with time: during the early stages of breakdown of wood there is slow decay of cellulose and lignin but the rate of decay may increase, and the resource quality improve temporarily, as the initial colonizers release more easily available carbon and nitrogen or are themselves decayed in secondary resource successions. On the other hand resource quality may deteriorate more uniformly with time; a soft sugary fruit may have abundant simple carbon sources at the beginning which are converted eventually into the less available compounds such as the remains of the chitinous walls of the decomposer fungi themselves.

The rate of decomposition is governed by environmental factors, for any one resource. Although temperature, water availability, nutrient availability (especially nitrogen and phosphorus) are usually important, the resource quality has an overriding effect.

Apart from these resource successions there are long term changes in vegetation as rocks are converted to soil and lakes fill up with sediment and eventually become climax forest. These long term changes are called seres or ecological successions. Each plant species seems to have a particular microbial community associated with it which varies from the next species either in the quantity of microbes or in the microbial species present. The microbial species differences for different plants are especially true of pathogens but do occur for saprotrophs as well. As plant species change in a sere the micro-organisms will change too. Micro-organisms themselves may be very important in seral changes, indeed they may be the only organisms at the beginning of the sere, e.g. lichens on bare rocks.

Resource successions and seral successions are basically different in many aspects. For example the availability of nutrients to heterotrophs changes in different directions, from low to high in a sere and the reverse for most resource successions. Some of these characteristics are summarized in Table 1.6 and the changes are further discussed by Swift, Heal & Anderson (1979).

Successions imply that different organisms are adapted to different conditions so, though there will be places where one or other organism is so well adapted as to be in sole command, there will be 'edges' in space and time where conditions are changing and one organism is taking over from another, i.e. there will be competition. Competition may be for several different resources at once (space, nutrients, light, etc.) and there are several different strategies for competing. Many microbes commonly isolated from plants have one or more of the following characteristics: they grow very fast, produce lots of spores, produce antibiotics, are tolerant of other organisms' toxins, are undemanding in their nutrient requirements and tolerate a wide range of environmental conditions. An organism with these abilities is equivalent to an '*r*' strategist in higher plant and animal ecology; it reproduces and grows fast so that it is ubiquitous and able to colonize new resources or to take quick advantage of favourable environmental conditions. For example many of the commonly-isolated fungi (*Aspergillus*, *Penicillium*, *Mucor*) have this high competitive ability and so do some bacteria (*Pseudomonas*, *Bacillus*). Conversely there are microbes that do not have the attributes listed above and they are equivalent to the '*K*' strategists. They avoid intense competition by possessing some attribute which gives them an advantage; it might be that they grow at very low nutrient levels or on nutrients which cannot be utilized by other organisms. In this way they occupy a unique niche and so do not have competition. Thus basidiomycetes and actinomycetes can degrade complex molecules such as lignin or humic material in plant litter. '*K*' strategists tend to be present and active more or less all the time and are not so subject to violent population fluctuations as the '*r*' strategists are. Plant pathogens do not fit easily into these categories; they utilize a nutrient source not available to others, the living plant, and once they have penetrated the host defenses they have nutrients and space within the plant at their disposal with very little competition from other microbes. In this sense they are '*K*' strategists, but some pathogens also have a very high reproductive rate and may have rapid changes in their population levels in an epidemic (which should strictly be called an epiphytotic when considering plant diseases).

In this chapter I have tried to introduce micro-organisms, the methods of handling them and the possible interactions between microbes themselves and between plants and microbes. In the next chapter we will consider in more detail the relationship between saprotrophs, pathogens and the plant, including the plant's defense mechanisms against microbial attack.

Selected references and further reading

Aaronson, S. (1970). *Experimental microbial ecology*. Academic Press, New York. pp. 236.

Ainsworth, G.C. & Sussman, A.S. (& Sparrow, F.K.) (1965–1973). *The fungi: an advanced treatise*. Volumes I–IV. Academic Press, New York.

Alexander, M. (1971). *Microbial ecology*. Wiley, New York. pp. 511.

Brock, T.D. (1979). *Biology of micro-organisms*, 3rd edition. Prentice-Hall, New Jersey. pp. 802.

Buchanan, R.E. & Gibbons, N.E. (Eds) (1974). *Bergey's manual of determinative bacteriology*, 8th edition. Williams & Wilkins, Baltimore. pp. 1268.

Burns, R.G. & Slater, J.H. (Eds) (1982). *Experimental microbial ecology*. Blackwell Scientific Publications, Oxford. pp. 683.

Campbell, R. (1983). *Microbial ecology*, 2nd edition. Blackwell Scientific Publications, Oxford. pp. 191.

Dawes, I.W. & Sutherland, I.W. (1976). *Microbial physiology*. Blackwell Scientific Publications, Oxford. pp. 185.

Gibbs, A. & Harrison, B. (1976). *Plant virology; the principles*. Edward Arnold, London. pp. 292.

Hawker, L.E. & Linton, A.H. (Eds) (1979). *Micro-organisms: function, form and environment*. Edward Arnold, London. pp. 391.

Johnson. L.F. & Curl, E.A. (1972). *Methods for research on the ecology of soil-borne plant pathogens*. Burgess, Minneapolis. pp. 247.

Margulis, L. & Schartz, K.V. (1982). *Five kingdoms, an illustrated guide to the phyla of life on earth*. Freeman, San Francisco. pp. 338.

Primrose, S.B. & Wardlaw, A.C. (Eds) (1982). *Sourcebook of experiments for teaching microbiology*. Academic Press, London. pp. 766.

Round, F.E. (1973). *The biology of the algae*, 2nd edition. Edward Arnold, London. pp. 278.

Round, F.E. (1981). *The ecology of algae*. Cambridge University Press, Cambridge. pp. 653.

Sleigh, M. (1973). *The biology of protozoa*. Edward Arnold, London. pp. 315.

Stanier, R.Y., Adelberg, E.A. & Ingraham, J.L. (1976). *General microbiology*, 4th edition. MacMillan, London. pp. 871.

Starr, M.P., Stolp, H., Trüper, H.G., Barlows, A. & Schlegel, H.G. (1981). *The Prokaryotes: a handbook on habitats, isolation and identification of bacteria*. Springer-Verlag, Berlin. Vols 1 and 2. pp. 1–1102 + I1–156 and pp. 1103–2284 + I1–156.

Swift, M.J., Heal, O.W. & Anderson, J.M. (1979). *Decomposition in terrestrial ecosystems*. Blackwell Scientific Publications, Oxford. pp. 372.

Whitcomb, R.F. & Tully, J.G. (1979). *The mycoplasmas: Volume 3: plant and insect mycoplasmas*. Academic Press, New York. pp. 351.

Wicklow, D.T. & Carroll, G.C. (Eds) (1981). *The fungal community*. Marcel Dekker, New York. pp. 855.

2
Micro-organisms as Saprotrophs and Plant Pathogens

Classification of micro-organisms as saprotrophs and pathogens

Despite the very wide range of micro-organisms discussed in the last chapter, those associated with plants are mostly necrotrophs or saprotrophs. A few have been called obligate pathogens, living and actively growing only on or in living plants and not able to grow in artificial culture. There are many facultative pathogens able to parasitize plants but also capable of saprotrophic growth on dead organic matter and laboratory media. This system of classification confuses two different criteria, firstly whether in nature the organisms grow as pathogens or not and secondly whether they can be grown in culture. The latter is not a good character for it depends on the state of knowledge: some rusts can be grown in culture and some cannot, but that does not make them fundamentally different, it merely reflects our ignorance of how to grow them. The division of pathogens into obligate and facultative based on their ability to grow in culture is therefore now being replaced.

Some microbes are clearly dependent on living plants for all the main stages in their life history, their nutrition depends on living cells; these are biotrophs. Other organisms can invade plants, but they kill them in the process and always live off dead tissue: these organisms are necrotrophs. There are also microbes that can never overcome plant host defense systems and always live on dead material that they have not themselves killed: these are saprotrophs and their nutrition is saprotrophic.

Even these criteria are not clear-cut for there are microbes which are usually saprotrophic but, in the presence of a weak host, they may be able to overcome the limited resistance and therefore become pathogens. For example, senescent tissue or plants already suffering from one or more diseases may be parasitized by organisms which are usually saprotrophs.

These various sets of terminology have been brought together by Lewis (Lewis, 1973) who devised a scheme which has the advantage of being comprehensive enough to cover all nutritional types of heterotrophic micro-organisms. There are five main groups: (1) obligate saprotrophs, (2) facultative necrotrophs, (3) obligate necrotrophs, (4) facultative biotrophs, and (5) obligate biotrophs. These groups are based on symbiotic organisms in the original broad sense of de Bary, including all organisms living together, both parasitic and mutualistic organisms (p. 14). No attention is given to the behaviour in artificial culture, the 'obligate' and 'facultative' referring to their behaviour in nature.

Obligate saprotrophs are organisms, living on the plant or in the surrounding soil, which are never pathogenic and which never form mutualistic symboises. This is by far the largest group and includes the litter decay organisms, most of the normal soil microbial population and epiphytic microbes on stems, leaves and roots. Some consider the relationship with roots to be a loose form of symbiosis since the organisms receive nutrients from the roots but there is no evidence that these organisms are anything but normal saprotrophs taking advantage of a readily available, if rather specialized, form of plant litter (but see *Azospirillum*, p. 124). The position of some bacteria and fungi on senescing leaves and fruits may also pose problems: when has the leaf died of its own accord and been colonized by saprotrophs and when has its dying been helped along a little? These difficult cases apart, there are a great number of microbes that are clearly obligate saprotrophs.

Facultative necrotrophs are unspecialized pathogens (Garrett, 1970) or opportunistic organisms that can live quite well as saprotrophs but may attack weakened hosts as the opportunity arises or may attack hosts as one part of their life cycle. Thus *Pythium* and *Rhizoctonia* cause damping-off of young seedlings but not of mature plants. *Penicillium* and *Rhizopus* can cause fruit rots of apple if there is a wound but will not normally invade healthy fruits. *Armillaria mellea* is a virulent pathogen of trees and shrubs, causing group dying in plantations and gardens but once it has killed a tree it lives on the roots and in the stump as a saprotroph, using it as a food source to support growth through the soil or along roots to spread to nearby trees (p.144). Lewis did not include bacteria in his classification but some bacterial wilts (e.g. *Pseudomonas solonacearum*) and soft rots of stored vegetables (*Erwinia*) probably should be included, in that they occur commonly in soil and gain entry through wounds caused by mechanical handling or by other pathogens.

The boundary between the facultative and the *obligate necrotrophs* is not well defined. Obligate necrotrophs spend most of their lives as pathogens and it is in this phase that they reproduce and spread. They kill the plant before using it for food and they may exist or survive saprotrophically but they have a very low competitive ability and do not, in natural conditions, grow extensively as saprotrophs. The vascular wilts caused by fungi (*Verticillium, Fusarium*) are in this group as is the take-all fungus (*Gaeumannomyces graminis*). The latter grows actively on roots of cereals and grasses, killing them and possibly the plant, and it merely survives in the soil on rubbish, dead roots, etc., left after harvesting. Unless a new cereal crop is planted the fungus will eventually die out: it cannot survive for long periods as a saprotroph. *Erwinia amylovora*, which causes fire blight on fruit trees, is an example of a bacterial obligate necrotroph: it only grows on live hosts, though it exists saprotrophically for dissemination by insects (especially bees) in soil and on twig surfaces and roots.

Facultative biotrophs are fungi that require the living plant for most of their life history, though they naturally have an independent free-living saprotrophic stage as well. Biotrophs do not immediately kill the plant in the process of using it as a food source. Facultative biotrophs include mutualistic relationships such as the basidiomycete mycorrhizal fungi, some of which can exist independently as litter fungi, though they do not have a high competitive ability. Some lichen fungi are also known to occur free-living. There are also some mutualistic bacteria such as *Rhizobium* and the alder endophyte *Frankia* which occur in the soil and in the living plant.

Finally there are *obligate biotrophs* which occur exclusively on living hosts and do not naturally occur as saprotrophs. The most important pathogenic fungi in this group are the rusts and smuts (Uredinales and Ustilaginales), powdery mildews (Erysiphales), *Plasmodiophora* (club root of crucifers) and *Claviceps purpurea* (ergot). Mutualistic obligate biotrophs include some of the lichens and most vesicular arbuscular and sheathing mycorrhizas (p. 131) which do not exist actively free-living in the soil: these have a great effect on the nutrition of ecologically- and economically-important plants. Some pathogenic bacteria are obligate biotrophs, such as *Agrobacterium tumifaciens* which causes crown gall on many plants at the point where the stem emerges from the soil. Obligate biotrophs are usually considered to be 'better adapted' to their host and therefore 'more advanced' in evolutionary terms, though there is little real evidence for this, other than that it seems better to have a food supply from a still living host rather than having to move on after killing the first food source.

Infection by obligate biotrophs can be affected by environmental factors. Thus strong light levels in the environment and high sugar levels in the plant may increase the infection by mycorrhizas and biotrophic rusts and smuts. Alternatively, lower light and low sugar concentrations may favour necrotrophs, such as vascular wilts, or make normal biotrophs like *Rhizobium* become necrotrophs as the root nodule degenerates. Mutualistic lichen symbioses may also break down under low light intensity as the algae degenerate. So, like all classification systems, there are borderline cases. The main advantages of the system are that it covers symbiotic organisms of all types and it does away with the artificial dependence on the ability to be cultured which varies with the skill of the investigator rather than being a fundamental characteristic of the organisms.

Virulence and resistance

The above discussion has raised the problems of alternating parasitic and saprotrophic life styles in one organism and of variable pathogen virulence. Why does an organism behave as a saprotroph under some circumstances and a pathogen under different ones? Why will an obligate biotroph infect one cultivar of a host but not another? This is complicated by the fact that not only is there a variation in the virulence of the pathogen but the host also may have variable resistance to the disease-causing organism. There is thus a spectrum from complete virulence and complete susceptibility (and so maximum disease) to avirulence and resistance (and so minimum disease or even none at all). There are obviously several interrelated phenomena here which need to be carefully sorted out.

Virulence is the infective capacity of individual strains of the pathogen as observed by the symptoms produced on the host. Resistance is the ability of the host to suppress or retard the activity of a pathogenic organism or virus. Virulence and resistance interact as noted above and so have to be qualified in relation to each other. If there is a single host and one organism causes a disease and another does not then the host is resistant to one but not to the other. Conversely an organism may be virulent on one host but not on another: whether the organism is virulent depends on which host or host cultivar it is tested on. If a host is susceptible then it is possible for a particular pathogen to infect, but the host may or may not be resistant to a particular race of that pathogen (Fig. 2.1).

Apart from the innate genetic characters of an organism it is also important just how much of an organism is present (the amount of inoculum) and also how vigorous that inoculum is. These two factors have been described together in the term 'inoculum potential' which has been defined as the energy for growth of a fungal pathogen available for infection (Garrett, 1970).

The inoculum may be widely dispersed in the soil or in the environment in general so that almost anywhere a seed is planted there is the risk of disease if the conditions favour it: such would be the case in the damping-off of seedlings. Some soil-borne diseases are not however so widely dispersed, only occurring in particular fields or in particular soil types where a susceptible crop has been grown previously. This is very common in many agriculturally-important diseases where the pathogen, usually a facultative or obligate

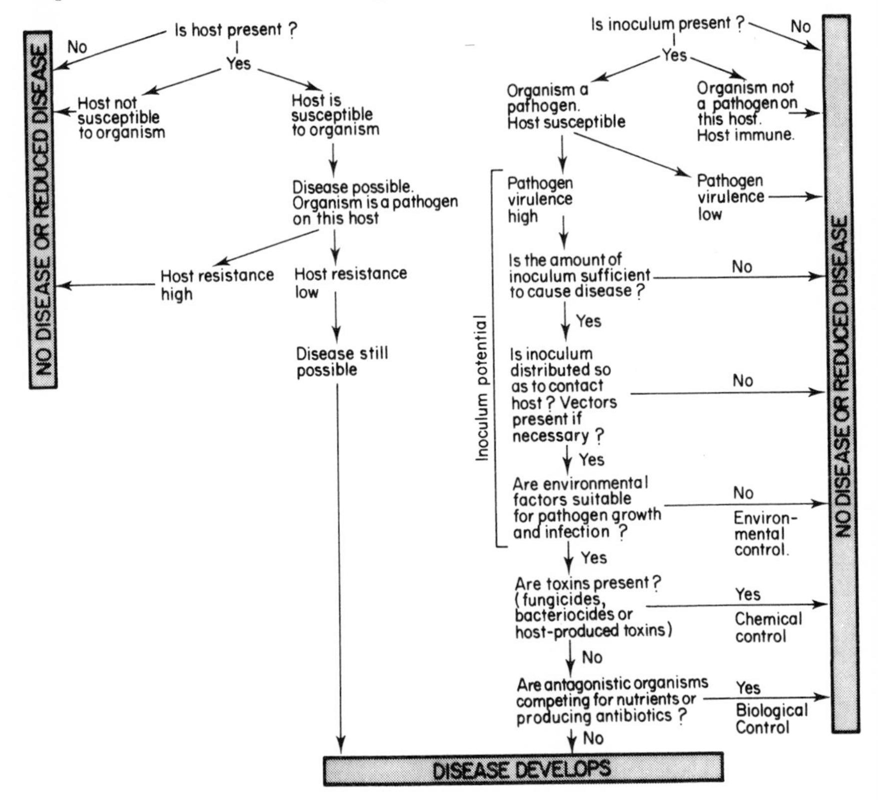

Fig. 2.1 The process of disease development looked at from the point of view of the potential pathogen (left) and of the potential host (right).

necrotroph of roots, survives on the debris from the previous crop (e.g. take-all disease of cereals, many vascular wilts, *Sclerotium* rot of onions). Leaf diseases often result from air-borne spores and so depend on the presence of the pathogen on other, usually nearby, plants and on suitable climatic conditions for transfer and/or infection. In some diseases the pathogen is transmitted by the plant itself, on or in its seeds. Alternatively there may be a specialized vector like an aphid in the case of viruses. So the initial requirement of disease is that virulent inoculum is present and inoculum potential is high enough to cause infection: understanding the distribution of inoculum and its persistance in the prevailing conditions is essential for disease control. The inoculum potential frequently has to exceed a threshold value for disease to develop. Thus air-borne spores may cross large stretches of water or even oceans but by the end of their passage the inoculum potential is too low to cause disease for most have been killed by desiccation or ultra-violet light and those that are left must land on their host plant in sufficient numbers and have sufficient food reserves left to grow and establish an infection.

Given that all these factors are favourable, then the environmental factors have to be suitable before penetration can occur, and only after this do all the main host resistance factors come into play. The most important environmental factors are the temperature for growth of the pathogen and the host, the presence of nutrients for the initial pre-penetration growth of the pathogen, and the relative humidity should usually be high so that the hyphae or bacterial cells are not killed by desiccation before gaining entry (see Fig. 2.1). The inoculum potential is partly determined by the environmental factors, but whether this potential is realized depends on the host's resistance and on any artificial or biological control measures which may be taken.

One crucial part of disease control is therefore to monitor the levels of disease and the occurrence of inoculum in order to predict disease or at least to recognize that the potential for disease is there if the other factors are or become favourable. This study of the epidemiology of the disease is done as routine in most western agricultural systems (Scott & Bainbridge, 1978). It may be done by knowing enough about the environmental factors required to be able to recognize risk periods, as is the case with *Phytophthora infestans*, potato blight. Alternatively regular monitoring of crops, with special combinations of test cultivars known to be susceptible to the disease, is carried out to spot the first symptoms and to look for the arrival of new races of the pathogen or the appearance of diseases which were previously not important due perhaps to changing agricultural practices or to the loss of effectiveness in chemical control measures.

Having tried to define susceptibility, resistance and virulence we must now look a little more closely at the mechanisms by which they arise. It is easy to say that a saprotroph does not cause disease because the host is immune, or that if pathogen virulence is low and the host resistance is high there will be no disease; but what makes a host immune or resistant? Why are not all organisms that occur on plant surfaces able to cause disease? There are two stages in determining whether or not disease will develop. The first is whether the organism will penetrate the host (Fig. 2.2) and this may explain why potato blight never causes disease on wheat: the wheat is immune to *Phytophthora*.

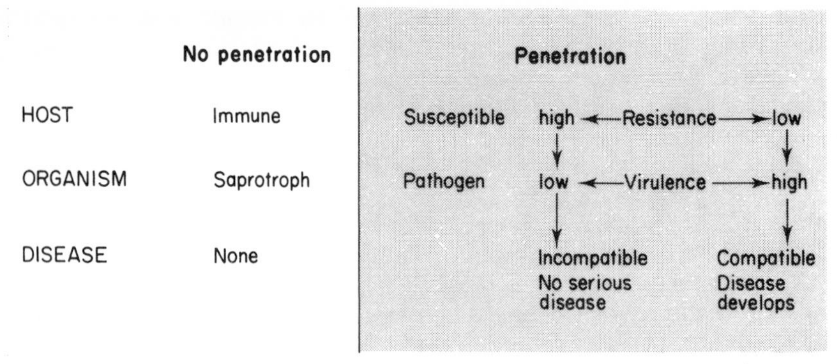

Fig. 2.2 Relationship between host immunity and resistance, and pathogen virulence. (Modified from Dickinson, C.H. and Lucas, J.A. (1982). *Plant pathology and plant pathogens*, 2nd edition. Blackwell Scientific Publications, Oxford. pp. 229.)

Immunity is an absolute characteristic of a species, not varying quantitatively. Having penetrated, there are qualitative differences in the virulence or resistance, and these will explain why one particular cultivar develops more serious symptoms than another when both are attacked by the same race of the pathogen. Quite a lot is known about this latter reaction but it is still very difficult to account in physiological or biochemical terms for the distinction between pathogens in general and absolute non-pathogens on a particular host. Saprotrophs often seem to have all the necessary enzymes, etc., to enable penetration to take place, yet it does not occur.

The distinction between saprotrophs and pathogens may take place before penetration, or even before spore germination, and various possible ways of recognition at such an early stage have been suggested, but not proved. Maybe there are deficiences in the nutrients available on some plants or alternatively there may be toxins present in the cuticle or in the surface waxes of leaves. Antagonistic saprotrophic micro-organisms on the plant surfaces could deter potential pathogens and cases are known where the host range of an organism is changed by removing or changing the normal surface flora. In the case of viruses it may well be the vector which is specific and artificial inoculations can produce infections which do not normally occur, simply because the virus is not transported there, the potential vector does not feed on that plant or the virus cannot get in if it arrives.

Mechanical barriers to infection

Given that a plant is susceptible to a particular micro-organism there are varying degrees of resistance which are genetically determined and which are manipulated by plant breeders to produce improved cultivars. This improved resistance may be based on (1) mechanical barriers which prevent or reduce infection and spread or (2) chemical barriers to inhibit or kill the pathogen. Particular examples will be discussed in later chapters but in general mechanical barriers may be formed before infection, such as the leaf cuticle, bark and epidermal cells. Plants may also initiate or re-activate meristems after infection to wall-off the invasion with cork cells. Individual cells may also

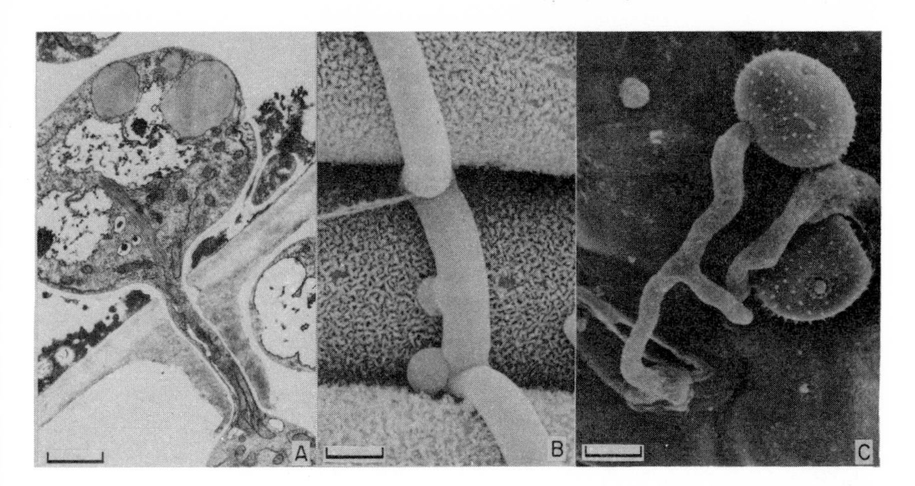

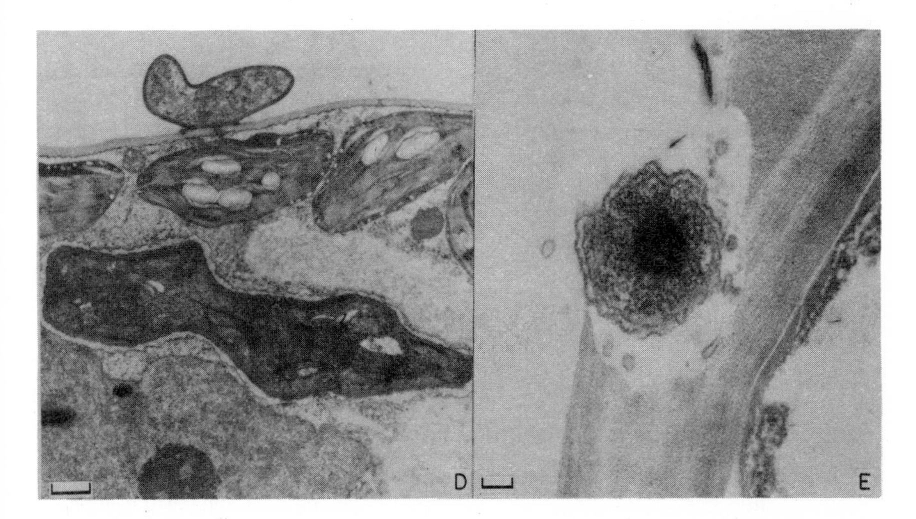

Fig. 2.3 Ultrastructure of microbial–plant interactions. **A.** Lignituber, in a wheat root, that has been ineffective in preventing the penetration and growth of the hypha. Bar = 5 μm. (From Faull, J.L. and Campbell, R. (1979). Ultrastructure of the interaction between the take-all fungus and antagonistic bacteria. *Canadian Journal of Botany,* **57**, 1800–8.) **B.** Appressoria of *Erysiphe graminis* on the surface of a barley leaf. Notice the hyphae of the fungus closely attached to the leaf surface, and the swollen appressoria at the penetration points. Bar = 5 μm. (Photograph courtesy of A. Beckett, Department of Botany, University of Bristol.) **C.** Urediospores of *Uromyces viciae-fabae* with germ tubes and appressoria. Bar = 5 μm. (Photograph courtesy of A. Beckett, Department of Botany, University of Bristol.) **D.** A haustorium of *U. viciae-fabae* in the leaf of bean. The fungus has penetrated the wall but not the host plasmalemma. The extra-haustorial matrix is electron-transarent and the fungal wall is dark. Bar = 1 μm. (Photograph courtesy of A. Beckett, Department of Botany, University of Bristol.) **E.** A bacterium penetrating the cortical cell of a root, digesting the wall as it makes a hole. Bar = 0.1 μm. (From Faull and Campbell, see A.)

respond and produce localized thickening of the walls around the invading organism. These callose deposits or lignitubers may prevent entry by some fungi, but others can pass through them and continue to be pathogenic (Fig. 2.3 A).

Fungi have an advantage over other microbial pathogens in that they grow at their tips: this means that if they are attached to a leaf surface by ordinary hyphae or by specialized structures called appressoria (Fig. 2.3 B and C) they can push their way into the plant through the cuticle and the cell wall. These simple structures do not usually stop fungi entering plants; cork layers and bark are however very effective at limiting entry and reducing spread and they often need to be broken by a wound before the pathogen can enter.

Fungi not only penetrate the cuticle but also grow within the host, either intercellularly or intracellularly. Necrotrophs kill the cells as they advance but biotrophs often form special interaction zones with the host. Usually the cell wall is penetrated and the host plasmalemma becomes invaginated to surround the developing specialized hypha of the fungus. This hypha, the haustorium (Fig. 2.3 D), has structures which seal it to the host cell wall. The host and fungal plasmalemmas have an altered structure and are separated by the fungal cell wall and by the extra-haustorial matrix. Nutrients pass across the haustorium–host interface into the fungus.

Bacteria are very different in their growth form to fungi, being basically single cells dividing by fission. They cannot therefore force an entry even through the cuticle; cell walls are also an effective barrier, though some bacteria can enzymically degrade the walls or pit membranes (Fig. 2.3 E). Bacteria usually enter plants through wounds or natural holes such as stomata and lenticels and grow into the intercellular spaces or in dead, empty cells with broken walls.

Viruses are again different. They are unable to multiply on the surface or penetrate plant cells on their own. They either enter through wounds, caused by pruning and during agricultural operations or they are injected into cells by the sucking mouth parts of their vectors (see Table 1.2).

Pre-formed chemical barriers to infection

If penetration occurs, and there are no post-infection mechanical barriers formed, then the resistance of the host depends on chemical means of defence. Again these may be pre-formed in the healthy plant or they may be produced in response to infection. Pre-formed inhibitors include the toxins in waxes, etc., which have already been mentioned and also chemicals which occur in the cytoplasm and cell walls. Phenolic compounds seem to predominate: tannins are common in bark and catechol and chlorogenic acid (Fig. 2.4) are found regularly. In potatoes for example such compounds are responsible for resistance to *Streptomyces scabies*. Saponins and unsaturated lactones are also known to be preformed inhibitors. Avenacin occurs in oats and is a major factor in preventing infection by root rots such as *Gaeumannomyces graminis* though some strains (*G. graminis* var. *avenae*) and other fungi such as *Fusarium avenaceum* contain enzymes which inactivate avenacin and are therefore able to be pathogenic on oats.

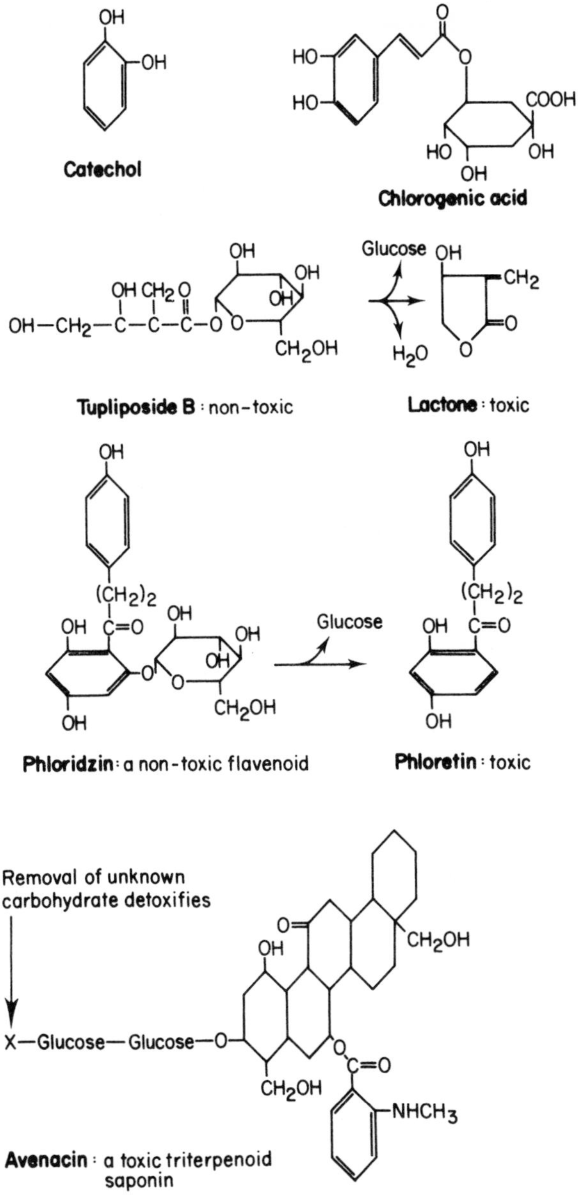
Catechol
Chlorogenic acid
Glucose
H_2O
Tupliposide B : non-toxic
Lactone : toxic
Glucose
Phloridzin : a non-toxic flavenoid
Phloretin : toxic
Removal of unknown carbohydrate detoxifies
X—Glucose—Glucose—O
Avenacin : a toxic triterpenoid saponin

Fig. 2.4 Chemicals occurring in plants which give some protection from attack by microbes.

This system is turned around in tulips (which have tuliposides, Fig. 2.4) and apples (with phloridzin, Fig. 2.4). In these cases the fungal enzymes may remove glucose and/or cause hydrolysis of the preformed substances and this makes the previously harmless compounds toxic. The pathogen has caused the activation of toxins which will prevent it from continuing the infection.

A further example of preformed chemical changes is found in cell walls. Pathogens often macerate tissues, and so kill the plant, by attacking the methyl galacturonic acid polymers in the middie lamella, causing the cells to fall apart. Many enzymes are involved but one of the first is pectin methyl esterase which removes the methyl groups from the methyl polygalacturonic acid, so allowing the polymer itself to be degraded by polygalacturonase. One defence mechanism, found in many cell walls, is to demethylate and form divalent cation bridges, especially with Ca^{2+}, and this gives protection against maceration.

Changes in host metabolism

One of the most obvious changes in the metabolism of the host following infection by fungi, viruses and bacteria is an increase in the permeability of the plasmalemma. Looked at teliologically this is very useful for it often provides the pathogen with food. The next obvious change is a rise in the host respiration rate which is especially due to the increase in the pentose phosphate pathway (Bailey & Deverall, 1983). This produces 4- and 5-carbon sugars and uses the cofactor $NADP^+$. Syntheses based on the tricarboxylic acid cycle (TCA) cycle also increase including that of lipid from acetyl-CoA, proteins from α-ketoglutarate and other TCA cycle intermediates, the synthesis of nucleic acid, the synthesis of isoprenoids and terpenoids from the mevalonate-CoA pathway and finally phenols and related aromatic compounds from the shikimic acid pathway which starts in the pentose phosphate pathway (Fig. 2.5). These syntheses often use NADPH and ATP and produce $NADP^+$ and ADP. In later stages of infection there may be some uncoupling of oxidative phosphorylation so that ATP is no longer produced.

The reduction in ATP by the syntheses and by the possible uncoupling results in the overall accumulation of ADP. There is also accumulation of oxidized cofactor, especially $NADP^+$. These two factors are the main control points for the Embden–Meyerhoff–Parnas glycolysis and the pentose pathways (Fig. 2.5), the rates of which therefore increase further and so give the observed rise in the respiration rate.

The products of these syntheses are interpreted first of all as the messenger ribonucleic acid and the proteins for the production of the new enzymes to initiate the increased activity. Many of the products of the syntheses such as phenols, flavonoids, isoprenoids and terpenoids (Fig. 2.5) have antifungal and sometimes antibacterial activity. They are produced as a result of infection and are called phytoalexins: they feature prominantly in discussions of host resistance. Some say that they have nothing to do with defence against primary pathogens but that they may be useful in preventing secondary infections of lesions. Most pathologists however maintain that they accumulate at the right time, in the right places and in sufficient quantity to be of great importance in the control of primary infections in many plants. Phytoalexins (Fig. 2.6) are host specific but they may be induced by many different biotic and abiotic stimuli. Wounding, toxic chemicals, pathogens and sometimes their culture filtrates will all induce phytoalexins in the plant and each plant species always produces the same limited number of them which are different to those produced by other hosts in response to the same stimuli (Bailey & Mansfield, 1982).

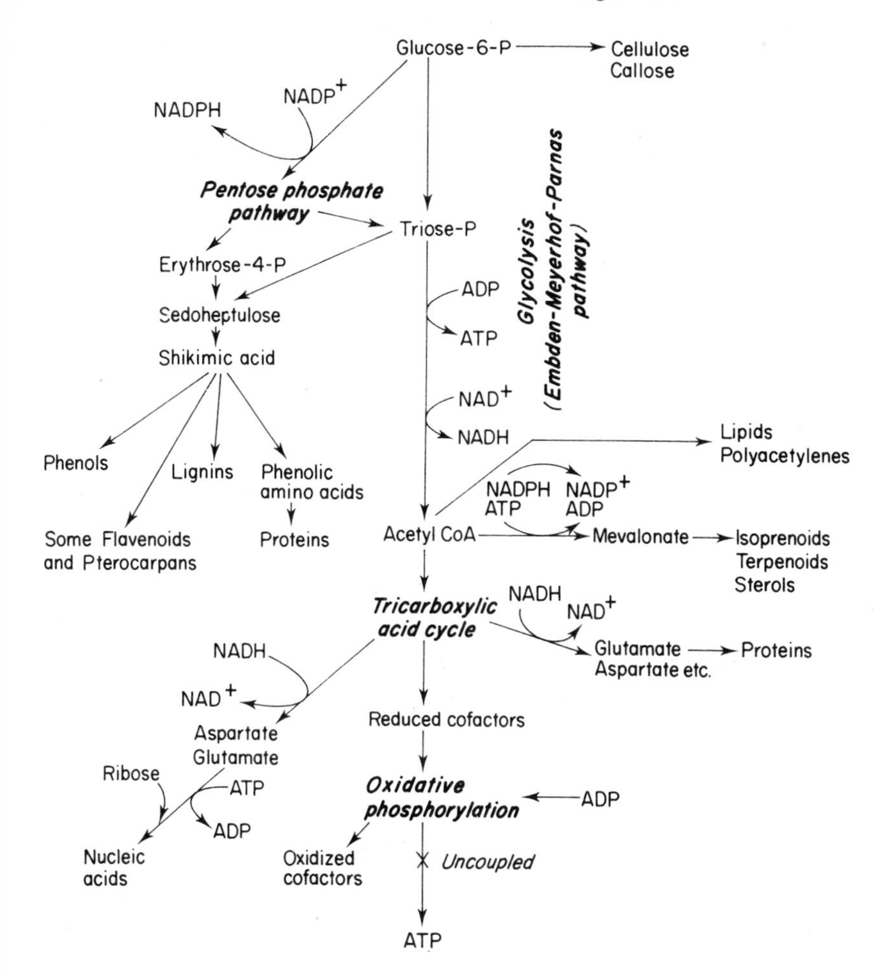

Fig. 2.5 Outline of biochemical pathways involved in the host response to pathogens.

Most of the work on phytoalexins has been done with fungi, but where bacterial diseases have been investigated (e.g. *Erwinia* on potatoes and *Pseudomonas* on bean leaves) it has been found that phytoalexins are produced and some of them at least are effective against bacteria. They are produced in lesions formed by viruses but they do not appear to be inhibitory to the virus multiplication and are probably just a wound response. Most of the phytoalexins that have been studied to any great extent are from the plant families Leguminoseae and Solanaceae. Very few have been found until recently in some important familes such as the Gramineae.

Phytoalexins have been shown to be induced by specific substances produced by the disease agent or by the host in response to wounding. Such substances are called elicitors and they may be one of the systems of recognition and specificity involved in the host–pathogen interaction.

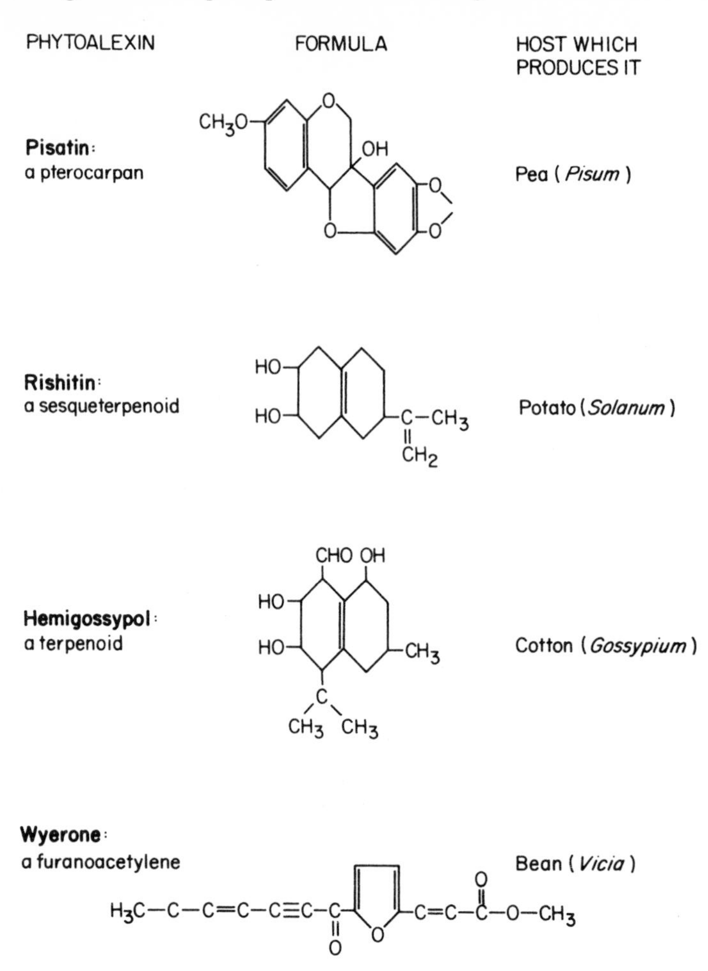

Fig. 2.6 Some examples of phytoalexins important in defence mechanism of plants attacked by micro-organisms.

The hypersensitive response

The biochemical changes discussed above are reflected in the visible symptoms of the disease. Thus if the physical and chemical defences fail there is a spreading lesion. Even if the defences work there will often be a few cells killed before the spread of the infection is limited (Fig. 2.7, column A). The rapid death of a few cells around the point of infection is known as the hypersensitive response (Fig. 2.8). This in itself may limit the growth of obligate biotrophs but, in theory anyway, facultative biotrophs and necrotrophs should be able to grow through the dead cells. They may not do so because phytoalexins accumulate in the dead cells, though the phytoalexins are probably not

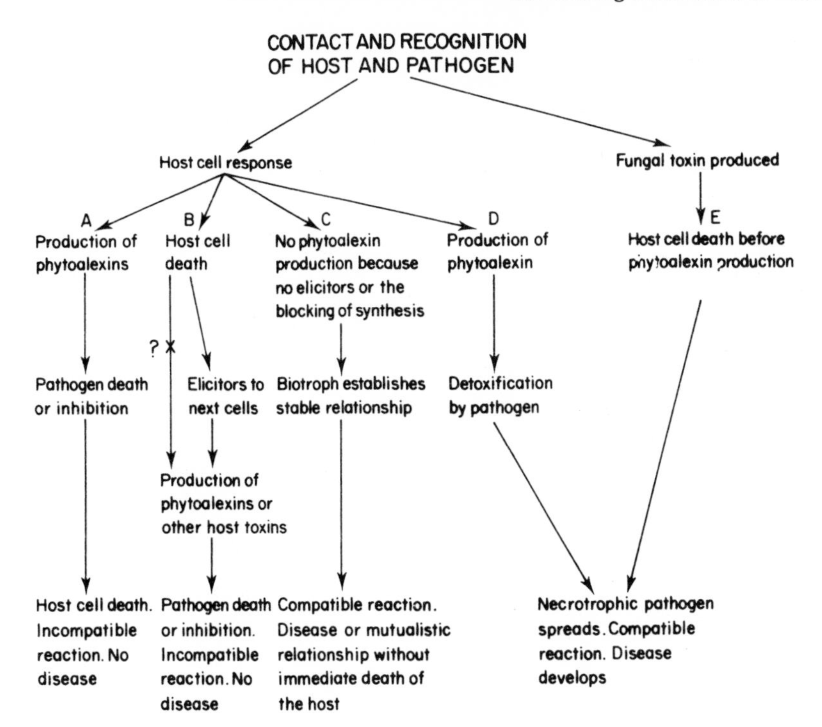

Fig. 2.7 Sequence of possible events during infection, with alternative explanations of the place of cell death, the hypersensitive response and avoidance of host response in the overall reaction to the pathogen.

produced by the cells before their die, they diffuse in from the surrounding live cells which may be triggered by some product of the few dead cells (Fig. 2.7, column B). It seems clear that, though related, the production of phytoalexins and the death of cells are not usually cause and effect. It could be that death results in phytoalexins or that they occur in such quantities as to kill the host as well as the pathogen; however these two options are unlikely on present evidence.

Both bacteria and viruses can produce the hypersensitive response, but it seems unlikely that it stops bacterial spread which is mostly extracellular.

Overcoming host resistance

Virulent pathogens, those that cause disease even on cultivars that have some resistance to infection, are able to overcome host resistance. Obligate biotrophs do not cause necrosis and death of the plant organ in the short term (Fig. 2.7, column C), indeed they may cause an increase in photosynthesis, the accumulation of food reserves and the formation of 'green islands' in an otherwise senescing leaf. The perpetuation of these stable relationships is difficult to explain on the basis of present mechanisms. They are similar to the

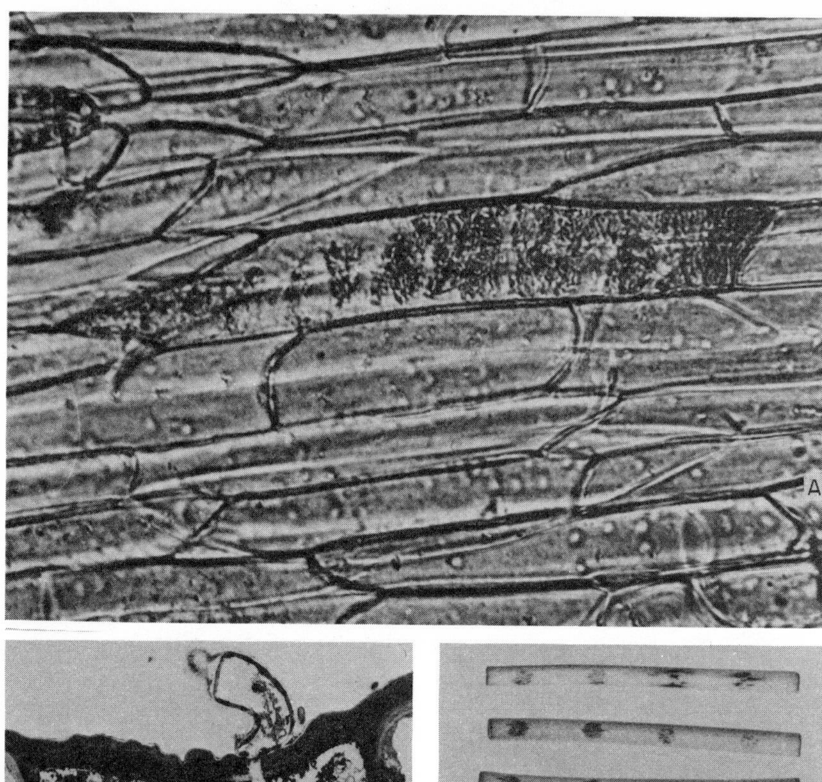

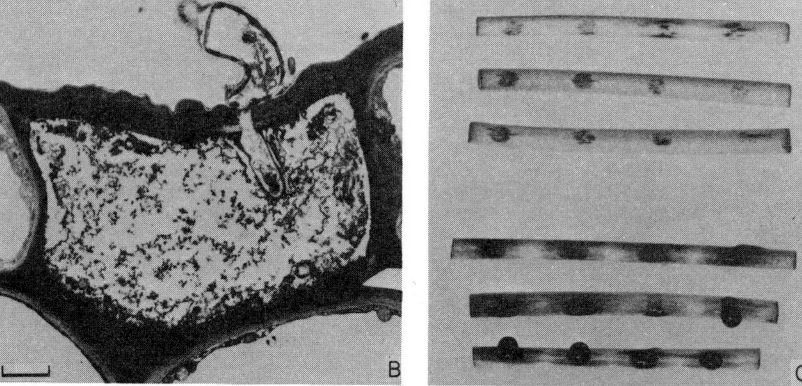

Fig. 2.8 The hypersensitive response in *Phaseolus* infected with *Colletotrichum lindemuthianum*. **A.** The death of a single cell: the granular cytoplasm is typical of the hypersensitive response. Surface view: light micrograph. **B.** A similar cell, seen in transverse section, with penetration by the potential pathogen which has died as the cell underwent the hypersensitive response. Surrounding cells are healthy. Bar = 5 μm. Electron micrograph. **C.** Hypocotyls infected with *C. lindemuthianum*. In the top three there has been the hypersensitive response and no disease development. In the bottom three hypocotyls the disease has established itself with spreading lesions. (Photographs courtesy of R.J. O'Connell and J.A. Bailey, Long Ashton Research Station, University of Bristol.)

other long term symbioses such as mutualism, in which the system of balance is also not understood.

Obviously virulent pathogens must possess the cell wall degrading enzymes to penetrate the host, even in the presence of lignin and callose deposition.

However many saprotrophs also have these enzymes, so what is special about the pathogens? The answer probably lies mostly in the biochemical host defences just described, and there are various strategies for overcoming this host resistance. Firstly the fungus may produce no elicitors and so will not induce the production of phytoalexins (Fig. 2.7, column C). Secondly the potential phytoalexin production can be blocked by pathogen toxins (Fig. 2.7, column C) acting either at the gene transcription stage or on the enzymes in the pathway essential for phytoalexin production. Thirdly the phytoalexins may be produced but they may not be toxic to the organisms (Fig. 2.7, column D); for example *Botrytis fabae* can degrade broad bean (lima bean) phytoalexins but *B. cinerea* cannot. *B. fabae* is a pathogen causing chocolate spot, *B. cinerea* is a saprotroph. Finally the fungus, a necrotroph, may produce a toxin which specifically kills the host cells so quickly that there is no response (Fig. 2.7, column E).

There are also various other ways of avoiding the host response by, for example, invading the dead tissues like the heartwood of tree trunks (p. 90). Other fungi grow quickly along the plant surface, only putting down feeding hyphae into the host at intervals. The feeding hyphae probably induce a response but by this time the surface growth is over a new part of the surface and is not affected by the response behind it (p. 143).

This chapter has attempted to outline very briefly the ways in which micro-organisms attack a host and the way in which that host may respond to prevent at least some of the attacks from being successful. The most important factor to remember is the variation in the virulence and the resistance which forms the continuous spectrum from the serious pathogen to the harmless saprotroph. Who does what to whom depends on the races and the cultivars present in the interactions and also on the presence of other microbes and on the environmental conditions. Microbes in nature do not occur in nice pure cultures under controlled environmental conditions. There are all sorts and conditions of micro-organisms present, and the colonization and growth on and in plants, which will be discussed in the following chapters, will depend on the outcome of the interactions between them as individuals or populations and between the microbial community and the plant.

Selected references and further reading

Bailey, J.A. & Mansfield, J.W. (1982). *Phytoalexins*. Blackie, Glasgow. pp. 334.

Bailey, J.A. & Deverall, B.J. (Eds) (1983). *The dynamics of host defense*. Academic Press, London. pp. 233.

Dickinson, C.H. & Lucas, J.A. (1982). *Plant pathology and plant pathogens*, 2nd edition. Blackwell Scientific Publications, Oxford. pp. 229.

Garrett, S.D. (1970). *Pathogenic root-infecting fungi*. Cambridge University Press, Cambridge. pp. 294.

Horsfall, J.G. & Cowling, E.B. (Eds) (1978). *Plant disease: an advanced treatise*. Volumes I–V. Academic Press, New York.

Lewis, D.H. (1973). Concepts in fungal nutrition and the origin of biotrophy. *Biological Review*, **48**, 261–78.

Scott, P.R. & Bainbridge, A. (1978). *Plant disease epidemiology*. Blackwell Scientific Publications, Oxford. pp. 329.

3
Microbiology of Flowers, Seeds and Fruit

This and the following chapters will consider the microbiology of particular plant parts and it seem reasonable to start with flowers and seeds, and their associated structures, from which the new generation of most land plants are derived. It should be stressed that the literature here is overwhelmingly biased towards agricultural crops and their fungal disease and spoilage problems.

Flowers

There is a large range of flower structures, and there are major differences in microbial nutrient availability between the wind-pollinated flowers and those that use sugar-rich nectar to attract pollinating animals, especially bees. This variety will not worry us unduly because very few genera of flowering plants have in fact been investigated microbiologically, only apple and vine have been looked at in any great detail. Most of the information is on yeasts, in relation to later fruit colonization and the production of fermented drinks. Filamentous fungi are occasionally mentioned in the literature but bacteria have been reported very inadequately, though they must be present since they are also found in ovules and young unripened seeds (p. 42). The transient nature of most flowers must limit the degree of colonization by microbes, which arrive from the air flora or on the bodies of visiting insects.

The filamentous fungi reported are mostly common soil and plant surface saprotrophs such as *Cladosporium cladosporioides*, *Alternaria tenuis*, *Penicillium oxalicum* and *Fusarium* spp. There are also some pathogens, such as *Aschochyta cucuminis* which was found in a study of cucumber (Raymond *et al.*, 1959). It is probable that most of these fungi were casual soil contaminants and there is no evidence of saprotrophic growth in these or most other flowers.

Some necrotrophic pathogens attack flowers, the most common of which is *Botrytis cinerea*, the common grey mould, which rots buds and flowers of many plants especially at low temperatures and in the wet conditions of late autumn and winter.

Not all flowers have a yeast flora. The yeasts present on or in flowers are dependent on the nectar as a food source; though many are present on petals they are probably not actively growing there. The estimates vary between flower species and with the geographical location, but less than 40 to 50% of freshly-opened flowers have yeasts, though the number infected and the population present tends to increase, together with the other saprotrophs, as the flowers fade and the seeds begin to develop. As in the case of leaves, fruits and most other plant parts, the yeast flora of flowers is usually less under greenhouse conditions than in the open. Most of the flower yeasts are asporogenous and include *Candida, Torulopsis, Kloeckera, Rhodotorula* and *Aureobasidium pullulans*. There is not enough information to be certain, but it is apparent that different flowers may have different floras and also different numbers of yeasts present. Apples seem to have an especially rich flora (Fig. 3.1). Within any one flower the stamens have the greatest number of different species, though not necessarily the highest number of individuals. Particular species may only be found on one flower part.

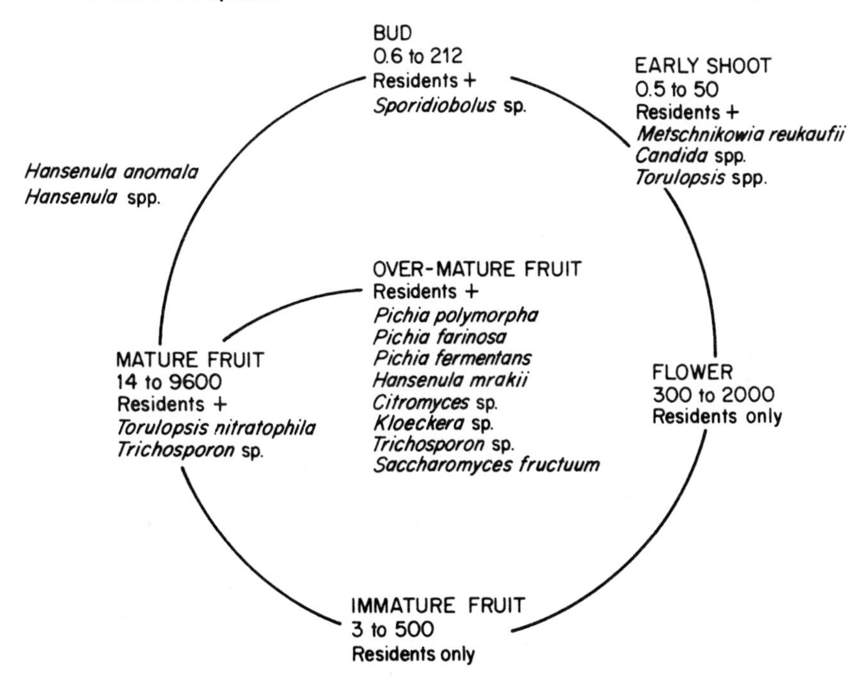

Fig. 3.1 Yeasts (numbers × 10^3 g^{-1}) associated with cider apple trees at various stages of development from dormant bud, early shoots, flowers, immature, mature and over-ripe or mummified fruit back to dormant bud. (Modified from Davenport, 1976.)

There are associations of yeasts with animals, especially insects such as bees, which may visit flowers. This association may be very specialized as in the case of the fig wasp (*Blastophaga psenes*) which is trapped within the hollow receptacle and lives there, ensuring pollination of the flowers: the wasp always carries *Candida guilliermondii* var. *carpophila*, though what this yeast does is not known.

Insect pathogens may also be transmitted through flowers and are therefore part of their flora even though they may not affect the flower. Bee pathogenic mycoplasmas have been found in many open flowers. They may be helical (spiroplasmas) or straight forms (p. 5) and they do not multiply in the flower, merely use it as a place to transfer from bee to bee. No mycoplasmas have yet been reported to cause diseases of flowers.

True bacteria also occur in flowers. They may be present in the buds or may be air-borne or splash dispersed to the flower. As with yeasts, the numbers increase with the age of the flowers, to reach several million bacteria per gram wet weight (e.g. from 5×10^6 g^{-1} in newly-opened flowers to 24×10^6 g^{-1} in fertilized flowers of soybean). *Erwinia amylovora*, the cause of fire blight on rosaceous hosts, is known to multiply in the nectaries of flowers from which it may infect the fruit, kill the flowers and fruit, or infect the fruit spur and eventually the rest of the branch. Similarly *Pseudomonas glycinae*, which causes a rot of soybeans, has been found in about 1% of newly-opened flowers and in about 33% of fertilized ones. It is transmitted from the flower to the developing pods.

Fleshy fruits

The microflora of developing and mature fruits has again been especially studied on those of commercial importance in wine and cider making with the emphasis on yeasts. There are of course, numerous pathogens and spoilage organisms on fruits but first let us consider the flora of apparently healthy fruits. Organisms may occur both inside and outside the fruit without any apparent damage; for example at least 20% of healthy tomatoes contained Gram-negative bacteria in their flesh. A low frequency of microbial infection has been reported for apples, the flesh being sterile unless wounded or invaded by pathogenic organisms, usually fungi. Numbers of bacteria and yeasts can be very high in rotten fruits as they utilize the nutrients made available by the primary fungal pathogen(s). The core of apples may contain micro-organisms derived either from the flower or by entry from the blossom-end of the fruit which remains open in some varieties. The numbers are however low, there may be only one organism per gram wet weight.

The surface population of apples is also low on young immature fruits (Fig. 3.1) and is dominated by lactic and acetic acid bacteria with few yeasts. Such yeasts as there are tend to be asporogenous with *Torulopsis*, *Sporobolomyces*, *Cryptococcus* and *Rhodotorula* being the main genera. *Saccharomyces* only occurs on ripe fruits together with *Kloeckera apiculata* and *Zygosaccharomyces*. The latter is important in that it is resistant to the sulphur dioxide used to sterilize the juice and it may therefore cause problems during the fermentation of the juice to cider.

The microbial populations of soft fruits such as raspberries and strawberries (though the latter are not true fruits, only swollen receptacles) seems to be dominated by the filamentous fungi such as *Cladosporium*, *Rhizopus* and *Alternaria* which are derived by splash dispersal from the soil or are wind-blown. Yeasts, which also occur in quite high levels on ripe fruits, include *Kloeckera apiculata*, *Aureobasidium pullulans* and *Cryptococcus laurentii* at 10^5 or 10^6 g^{-1} wet weight. Bacteria also occur in similar numbers. These saprotrophs mostly grow on the surface of the fruit though pathogens may invade the tissue. *Botrytis* particularly, attacks strawberries and other soft fruit and is a serious necrotroph. It may be controlled by spraying and formulations based on benlate or related systemic fungicides are very effective. However resistance has developed to some of these fungicides and most spray programmes now involve several different fungicides to try to overcome this problem. The fungicides are switched more frequently than the pathogen can develop tolerant strains so that the latter do not become dominant.

In some wine-making the growth of *Botrytis* on the fully ripe grape is encouraged, or at least permitted: when humid conditions which allow the growth of the fungus are followed by temperatures over 20°C, the grapes shrivel and lose water. Deep river valleys with high humidities may produce such conditions. The grape has a high sugar content and produces sweet white wines, called Beerenauslese in Germany if only a proportion of the infected grapes are included, and Trockenbeerenauslese if entirely made from the shrivelled grapes: such fruit may be sweet but it also has very little juice and the wines are therefore expensive.

There are many pathogens of apple fruits, most of them necrotrophic fungi. They are combatted by varietal resistance and extensive spray programmes. Apart from rots there are diseases like scab (*Venturia inequalis*), which, though disfiguring, are not otherwise harmful; such diseases do nevertheless reduce the selling price of the apple. The most common rot on the tree is brown rot caused by *Monilinia* (= *Sclerotinia*) *fructigena* or *M. laxa* or *M. fructicola*. These may infect the flowers or stems but usually the developing fruit is attacked and turns brown with the characteristic concentric rings of the imperfect sporing pustules (Fig. 3.2). The fruit dries out to a brown wrinkled 'mummy' which may stay on the tree for a while but eventually falls off. The fungus overwinters on the mummy on the ground (these mummies are also important overwintering sites for saprotrophic filamentous fungi and yeasts). Re-infection occurs in the spring. Many fruits are attacked including apple, pear, peach, apricot, plums, cherry and almond. The rot may only develop in storage and, combined with that on the tree, results in severe losses all over the world. Control methods involve the use of many fungicides, especially protectant ones to prevent spring infections and to give post-harvest control. Strains of the fungi are developing tolerance to the fungicides in widespread usage.

There are many other microbes which cause storage rots of these and other fruits which are economically very important. Losses of up to 20% of the crops are known though the average loss is about 12% for temperate agriculture. Losses in the tropics are higher, probably about 30%. Table 3.1 shows the post-harvest losses of some fruit in New York; there are differences due to the cultivar and to the point of origin of the produce. Standards of disease

Table 3.1 Post-harvest losses of fruit and vegetables sampled at wholesale, retail and consumer markets in New York, with an analysis of the causes of loss. (Modified from Harvey, 1978. Reproduced, with permission, from the *Annual Review of Phytopathology* 16. ©1978 Annual Reviews Inc.)

	Loss at indicated markets				Cause of loss		
Commodity	Wholesale (%)	Retail (%)	Consumer (%)	Total loss (%)	Mechanical injury (%)	Parasitic disease (%)	Nonparasitic disorder (%)
Apples, Red Delicious[a]	0.9	1.0	2.6	3.6	1.8	0.5	1.3
Apples, Red Delicious[b]	0.2	0.2	1.5	1.7	1.1	0.2	0.4
Cucumbers[c]	–	5.0	2.9	7.9	1.2	3.3	3.4
Grapes, Emperor[a]	–	5.5	–	–	4.2	0.4	0.9
Grapes, Thompson[a]	–	10.5	–	–	8.3	0.6	1.6
Lettuce, Iceberg[a]	4.1	4.6	7.1	11.7	5.8	2.7	3.2
Oranges, Navel[a]	1.9	1.9	2.3	4.2	0.8	3.1	0.3
Oranges, Valencia[c]	1.3	1.2	2.0	3.2	0.2	2.6	0.4
Peaches	2.3	4.5	8.1	12.6	6.4	6.2	–
Pears, Bartlett[a]	–	1.9	4.0	5.9	2.1	3.1	0.7
Pears, Bosc[a]	–	4.9	5.2	10.1	4.1	3.8	2.2
Pears, d'Anjou[a]	–	2.5	1.6	4.1	1.6	1.7	0.8
Peppers, Bell[c]	7.1	9.2	1.4	10.6	2.2	4.0	4.4
Potatoes, Katahdin[b]	1.3	–	3.6	4.9	2.5	1.4	1.0
Potatoes, White Rose[a]	1.1	–	3.2	4.3	1.5	2.4	0.4
Strawberries[a]	5.9	4.9	18.0	22.9	7.7	15.2	–
Sweet potatoes	–	5.7	9.4	15.1	1.7	9.2	4.2
Tomatoes, packaged	–	6.3	7.9	14.2	2.5	10.7	0.9
Tomatoes, bulk	–	6.7	4.7	11.4	2.7	7.6	1.1

[a] Product from western United States.
[b] Product from eastern United States.
[c] Product from southern United States.

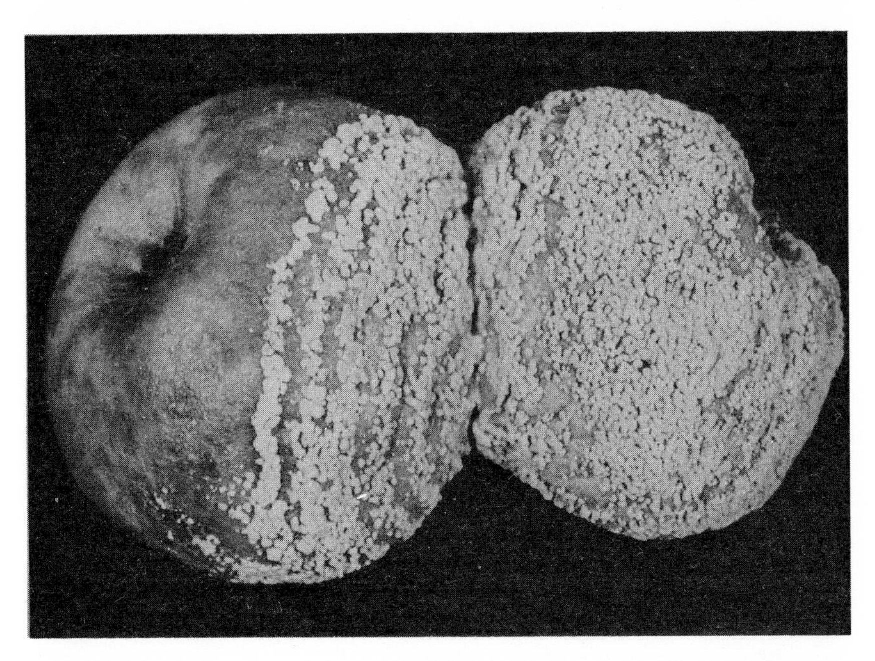

Fig. 3.2 Two apples infected by *Monilinia* showing concentric rings of imperfect spore pustules. (Photograph courtesy of R.J.W. Byrde, Long Ashton Research Station, University of Bristol. From Byrde, R.J.W. and Willetts, H.J. (1977). *The Brown rot fungi of fruit*. Pergamman Press, Oxford. pp. 171.)

prevention, handling and packaging are high in U.S.A. so these figures represent minimal losses and they do not include losses during growth of the crop. Some fruits, such as grapes, suffer particularly from mechanical damage but in others, e.g. oranges and tomatoes, microbial decay is more important. Most of the 'non-parasitic' disorders are caused by physiological problems arising during storage, at too low a temperature for example.

Most of the spoilage organisms are soil or air-borne spores from the field, harvesting or packaging machinery: there is always some damage during harvesting. Fungi are the main organisms in the post-harvest spoilage for many of the fruits are too acid for bacterial growth. There are characteristic fungal rots of different sorts of fruits, and even varietal differences but in general a few imperfect fungi such as *Penicillium*, *Aspergillus* and *Alternaria* are important along with some zygomycetes like *Mucor* and *Rhizopus*. These all produce many spores and are '*r*' strategist saprotrophs or necrotrophs present almost everywhere the fruit is grown and handled so that they take advantage of any damage or bruising and attack the fruit at all stages from harvest to consumption. The rots cause serious economic loss but are not usually a health hazard for the fruit is degraded so as to be unpalatable. This contrasts with contamination of high protein products where food-poisoning bacteria may occur without obvious spoilage of the food. There are some potential hazards from rotten fruit such as growth of *Penicillium expansum* which produces a toxin, patulin, but it is not usually serious on the fleshy fruits (see below for growth on seeds and nuts).

Many methods of control are practised to reduce the losses, the most important of which is care in handling to prevent damage and strict hygiene in the packing stations. Fungicides may be used pre-harvest to reduce inoculum on the fruit and in the washing water during preparation and packaging if this is appropriate for the fruit concerned. The control of environmental conditions during packaging and storage is also important but care is needed, for most fruit is picked under-ripe and matures during shipping and storage. Conditions which inhibit fungi are also likely to slow ripening. Low temperatures are used ($< 10°C$), though below 4°C chilling injury may occur to some fruits. An increased carbon dioxide level ($> 10\%$) slows rotting by fungi but has little effect on bacteria and yeasts. The relative humidity is generally kept rather high to prevent the fruit going soft but this also encourages fungal growth.

Hard fruits and seeds

We will consider together the true dry seeds and also things which are botanically fruits but which are dry and normally spoken of as seeds, such as cereal and grass fruits.

A great deal is known about the flora, especially the fungal flora, of seeds from the study of plant pathogens and from the routine testing for quality control (see Noble & Richardson, 1968; Baker, 1972). For example soybean carries up to 40 pathogen species, maize 43 and rice 50, as well as all the saprotrophs. Detailed listings for each commercial species of seed are available. Seed-borne pathogens have been known for a long time; smut fungi were first described as being seed-borne by Tillet in 1755, bacteria (*Xanthomonas phaseoli*) were reported in the 1890s and viruses (tobacco mosaic virus) were shown to occur on the seed surface by Allard in 1915. Organisms may be transported to the next generation of hosts either on the outside or inside of seeds, usually without any obvious sign of disease on the seed itself. Occasionally the whole 'seed' may be used for transport as in the case of *Claviceps purpurea*, ergot of grasses and cereal grains, where the infection results in the whole grain becoming a fungal resting structure called a sclerotium. This is dark brown or black and contains the ergot alkaloids which are responsible for the disease St. Antony's Fire or ergotism in the people who have eaten contiminated grain or flour derived from it. There are many symptoms of the disease depending on the dose which the person or animal receives but the main one is a restriction of the peripheral blood supply and the development of gangrene. Ergot alkaloids are used in medicine to reduce and control bleeding.

There are now seed certification schemes in many countries to prevent the transmission of plant diseases from generation to generation by seed-borne microbes. The detection rates may have to be very good to prevent transmission; in California a maximum of only 1 in 30 000 seeds is allowed to be infected with lettuce mosaic virus if the seed is sold on the commercial market. Seed transmission is one of the main ways in which viruses may re-infect the next generation of hosts (Table 1.2, p. 4).

The flora of cereal grains has been extensively studied, one paper lists over 270 organisms known from barley grains alone. Cereal grains are fruits and they therefore have several layers of tissues around the seed itself; microbes may grow on or between the palea and lemma as well as on the outer layers of the true seed. Table 3.2 gives a selection of organisms isolated from wheat and will serve as an example of cereals. The dominant organisms are the fungi *Alternaria*, *Fusarium*, *Epicoccum* and *Cladosporium* in the field. Most of these fungi seem to do little harm, though *Fusarium* may reduce germination. Bacteria occur on or in apparently healthy seeds, especially under the testa. The main genera are common soil or plant isolates such as *Pseudomonas*, *Bacillus subtilis*, *B. cereus*, *Alcaligenes*, *Corynebacterium*, *Erwinia*, *Aerobacter* and *Agrobacterium*. These organisms again seem to cause little harm; even though they occur within 5–15% of seeds examined the numbers are low, usually $< 100\ g^{-1}$ of seed. Surface numbers may be very much higher,

Table 3.2 Fungi present on cereal grains during growth harvest and storage. (Based on data of Pelhâte, 1968.)

	In field	Harvest	In storage
Common Isolated from 71 – 100% of samples	*Alternaria tenuissima* *Epicoccum nigrum* *Fusarium roseum*	*Cladosporium cladosporioides* *Aureobasidium pullulans* *Hyalodendron* sp. *Rhizopus nigricans*	*Penicillium viridicyclopium* *P. cyclopium* *P. spinulosum* *P. frequentans* *Aspergillus candidus* *A. echinulatus* *A. niger* *A. repens* *A. versicolor*
Frequent Isolated from 41 – 70% of samples	*Cladosporium herbarum* *Fusarium tricinctum*	*Acremoniella atra* *Chaetomium dolichotrichum* *C. globosum*	*Hemispora stellata* *Scopulariopsis brevicaulis* *Actinomyces griseus* *Aspergillus amstelodami* *A. flavus* *Penicillium viridicatum*
Uncommon Isolated from 11 – 40% of samples	*Septoria nodorum* *Fusarium nivale* *Botrytis cinerea* 2 other Deuteromycetes	*Verticillium malthousei* *Alternaria chartarum* *A. tenuis* *Sporobolomyces roseus* *Penicillium* spp. *Aspergilus* spp.	*Penicillium* 8 species *Aspergillus* 5 species Mucorales 4 species Actinomycetes 1 other Deuteromycete
Rare Isolated from 1 – 10% of samples	10 Deuteromycetes 1 smut	8 Deuteromycetes *Penicillium* sp. 2 Ascomycetes 1 Basidiomycete	7 Deuteromycetes *Penicillium* 10 species *Aspergillus* 14 species Mucorales 10 species

hundreds of thousands or even millions per g of seeds which is two or three orders of magnitude more than surface yeasts and filamentous fungi. In dry storage conditions the microbes are usually inactive though once sown in the soil they are of course wet and organisms will grow rapidly from the pre-existing inoculum, living on the exudates from the germinating seeds (see below). It has been suggested that saprotrophs aid germination by breaking down the hard pericarps and testas. Bacteria and fungi have also been shown to increase the germination of seeds by producing plant hormones (see below).

Other organisms which are reported to occur on seeds include yeasts, such as *Sporobolomyces* and *Rhodotorula* in numbers up to 10^6 g^{-1}, and myxomycetes (e.g. *Didymium*). The latter seem common though they are not extensively studied.

If the storage conditions are not ideal, especially if the seeds are wet, a special flora of storage organisms develops (Table 3.2). Fungi are dominated by *Penicillium* and *Aspergillus* which may cause loss of grain and also produce toxins (see below) and there are yeasts (e.g. *Candida*) growing within the seeds. Bacteria are very common under such conditions and, like the fungi, may be thermophilic since the temperature of bulk-stored seeds rises rapidly if rotting occurs.

Great attention has recently been paid to toxins produced by micro-organisms during storage. There have been many cases of disease or death of farm animals, and man, caused by eating seeds infected during storage. Fungi are mainly responsible since they will grow at lower water potentials than bacteria and so can cause problems in even slightly damp conditions. Furthermore seeds spoiled by bacteria are usually obviously rotted, but fungal growth can occur quite unobtrusively. The dangerous compounds produced are called mycotoxins. It should be made clear that this is a specialized problem of bulk-stored grains, nuts and oil seeds and not a general concern with food that has gone mouldy or with mould-ripened cheeses. Mycotoxins may cause human or animal diseases either by direct consumption of the product or by eating the milk or meat from animals which have fed on the affected food. The toxins are normally destroyed during refining of oils for human use, but products such as soy flour will retain the active toxin. In general, food and storage hygiene and quality control of the seeds or products as they are released from store, are good enough in the developed countries that there is no serious risk. Apart from direct poisoning there may be long-term problems with carcinogenic, teratogenic and oestrogenic effects.

There are many toxins involved but the main organisms are *Aspergillus* (especially *A. flavus*) *Penicillium* and *Fusarium*. As can be seen from Table 3.2 these are quite common on stored grains, and on other products, and they may be obligate or facultative thermophiles.

Toxins derived from the *A. flavus* group (11 or 12 species) are called aflatoxins (Fig. 3.3) and were first identified in 1963 though they have undoubtedly been around for as long as man and animals have stored food. The type of food stored is important; *A. flavus* growing on cereals does not produce much toxin but on nuts, especially peanuts, sunflower and cotton seeds (for oil production) there may be a serious problem. Levels of aflatoxin allowed in food vary from country to country but are about 20–100 p.p.b. The problem is further complicated by the fact that different species of animals vary in their susceptibility from LD_{50} of 0.3 to 18 mg kg^{-1} body weight. The main effect is on liver function, producing death at high dosage, but possibly also leading to liver cancer after prolonged exposure to even quite low levels. The most obvious examples have been in intensive poultry and pig-rearing farms where large numbers of animals have been fed with a batch of infected food.

Fusarium spp. produce several toxins. Zearalenone (Fig. 3.3) is formed by *F. roseum*, *F. tricinctum*, *F. oxysporum* and *F. moniliforme* especially when growing on maize. This toxin has oestrogenic properties resulting in infertility and small litter size in pigs and reduced egg laying in chickens. The tricothecenes (Fig. 3.3) are produced by *Fusarium* spp., *Trichoderma* spp. and *Stachybotrys* and they cause sores and haemorrhage, affect protein synthesis and membrane function and depress bone marrow activity.

Penicillium viridicatum and *Aspergillus ochraceus* may produce ochratoxins (Fig. 3.3) which cause haemorrhage and liver and kidney damage. *P. rubrum*, producing rubratoxins, may cause similar symptoms and also damage the central nervous system. Several *Penicillium* spp., including *P. expansum* (p. 41) may produce patulin (Fig. 3.3) when growing on fruits and cereals and the main symptoms are oedema, especially pulmonary oedema, and haemorrhage.

These are the major mycotoxins; there are many others which, though no less toxic, occur much less commonly because the fungi are less frequent on the stored seeds or the growth conditions are not suitable. The only method of

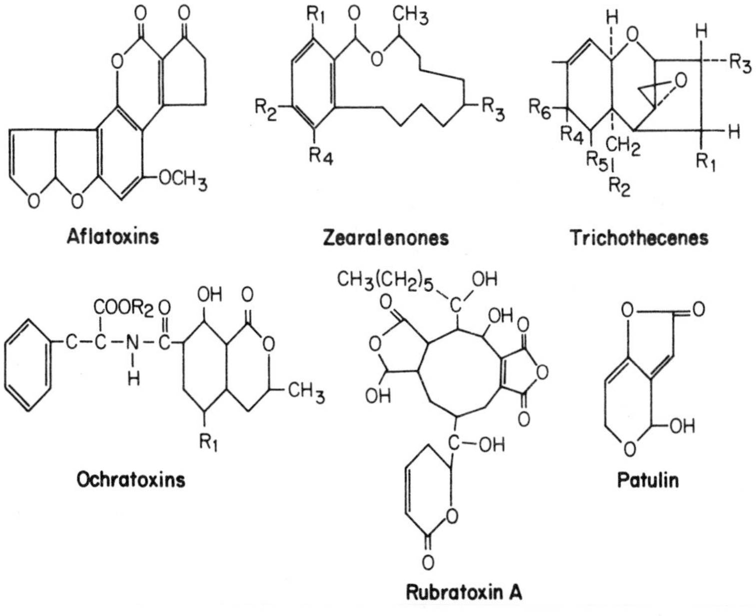

Aflatoxins **Zearalenones** **Trichothecenes**

Ochratoxins **Rubratoxin A** **Patulin**

Fig. 3.3 The structure of some mycotoxins produced by fungi growing on nuts and seeds.

controlling the production of toxins is by more careful storage, especially keeping the seeds dry with a relative humidity of less than 65–70%. At low humidities, with no free water, fungicides are not effective for they are not taken up by the microbes under these conditions. Hand sorting to remove damaged or mouldy kernels is also used, especially in the third world countries where labour is comparatively cheap. Avoiding sources of contamination, by not allowing nuts picked from trees to lie on the soil for example, is also important.

Seed germination microbiology

Microbial effects on germination may be beneficial such as the decay of hard pericarps noted above (p. 44) and this is obviously important for those species whose embryo cannot break out of the protective covering or cannot imbibe water through it. This is one of the reasons for stratifying some seeds. The seed, or fruit, is placed between layers of moist sand or soil and then left outside for some months, often over the winter. During this time the embryo may mature or the dormancy may be broken by moderate temperatures after chilling or freezing, but microbial weakening of the seed coat or the pericarp is also achieved. This is not done in large scale agriculture, though it is important in forestry and horticulture.

Soil and seed surface flora may also benefit the plant by producing various plant hormones. Many microbes, including common soil bacteria, produce extracellular products with plant hormone activity in culture. Some do so in nature, for example *Gibberella fujikuroi* is a fungus which causes a rice disease characterized by growth distortion because of gibberellin production. Production of hormones by saprotrophs has not been conclusively demonstrated in soil however. Inoculation of the seed with bacteria such as *Azotobacter*, *Clostridium*, *Bacillus*, *Pseudomonas* and *Arthrobacter* can produce increases in crop yeild (Brown, 1974). Many of these bacteria were used in the hope of increasing nitrogen fixation around the seeds and the developing plants; this is most probably not the cause of the increased yields for it is doubtful if there is enough available carbon for the fixation of significant quantities of nitrogen, a process which requires a lot of energy. Some of these seed treatments with bacteria increase germination rate and affect later root and shoot growth in ways similar to indole acetic acid and gibberellins. Protozoa may also produce plant hormones. There can be an effect on pathogens of the seed or plant, the inoculated bacteria contributing to biological control (see p. 150), though there is doubt about whether sufficient introduced organisms can become estabilished.

A related agricultural practice is the inoculation of seeds of legumes with strains of *Rhizobium* which will form more effective nitrogen-fixing nodules with that host (p. 130). This introduces the bacteria into the soil where the plant will grow and encourages later root nodule formation.

There are unfortunately a great many deleterious effects of micro-organisms on seed germination apart from these beneficial ones. The most obvious being the case of plant pathogenic microbes which attack the young emerging plants. This group of diseases is collectively known as damping-off and is all too common in both commercial nurseries and in amateur gardening. Several common soil fungi are involved, especially *Pythium*, *Fusarium* and *Rhizoctonia solani*. They are opportunistic, necrotrophic pathogens that attack weak seedlings or germinating seeds, especially in the cold conditions in early spring which reduce plant growth. Waterlogging of the soil may also be important, not only for the deleterious effect that it has on plants but also because some of the fungi (*Pythium*) need free water for their zoospores. Water and temperature stress increase the exudation of organic nutrients from the seed and this also encourages or permits the growth of microbes. Penetration may be through cracks in the seed coat or direct into the emerging root or hypocotyl. The disease affects a very wide variety of hosts, world wide, in almost all climates and soils. Two forms of damping-off are often distinguished, pre-emergent in which the seed rots and post-emergent where the germination is successful but the seedling is later attacked at or just below the soil level and the hypocotyl or stem rots and the plant falls over. Control is by the use of fungicidal seed dressings (e.g. organic mercury compounds) or sprays and by sterilizing the soil or composts in which the seeds are sown. The problem can be greatly reduced by not overwatering, by giving good drainage and by having suitable temperatures and air circulation. Recently biological controls have been used, and control agents are available commercially; usually the fungus *Trichoderma* is inoculated into the soil and kills or antagonizes the pathogens.

There are also a number of minor pathogens of seeds and seedlings which decrease germination and emergence. If seeds are treated with fungicidal dressings there is an increased number of seedlings even though there is no obvious disease in the absence of the fungicide. Fungi thought to be controlled include *Cylindrocarpon destructans*, *Fusarium sambucinum* and *Gliocladium roseum*. Bacteria, especially Gram-negative ones, have also been implicated in this reduced germination, particularly in wet conditions; seed dressings of aureomycin can improve germination though it is too expensive to use commercially. It is thought that the bacteria, and some of the fungi, may compete with the seed for oxygen.

These same sorts of problems have been important recently in commercial cereal growing using techniques of minimum tillage agriculture (p. 161). There are many different systems but basically no ploughing is done to mix the top soil. The surface may be rotavated or the soil tyned or sub-soiled or the seed may be sown direct into the stubble of the last crop using a special seed drill. These methods save time and fuel as they do not require so many passes over the field as do traditional methods and they are therefore cheaper and allow earlier sowing of winter cereals. Occasionally they permit the autumn sowing in a brief spell of fine weather when there would not be time for full cultivation. The disadvantages are some degree of soil compaction, especially on heavy soils, and the build-up of organic matter on the soil surface (p. 160). The increased decomposition on the soil surface results in anaerobic micro-habitats and the release of various toxins, including acetic acid. These anaerobic conditions result in stress on the germinating seedling which may be short of oxygen, suffering from acetic acid toxicity and possibly waterlogged due to soil compaction. This stress can increase exudation from the micropyle of the seed to up to 8 μm C per seed h^{-1} and this leads to increased growth of bacteria and fungi (especially *Gliocladium roseum* on barley) and a further decrease in the oxygen available to the seed. Quite extensive growth of fungi is possible (Fig. 3.4). Some success has been obtained by using bacterial inoculation to inhibit the fungi, but the bacteria themselves (*Azotobacter chroococcum*) may compete for oxygen. The most promising experimental treatment seems to be to coat the seed with a mixture of lime and calcium peroxide which neutralizes

Fig. 3.4 Barley grain infected by *Gliocladium roseum* which is growing especially around the micropyle where exudation is greatest. (Photograph courtesy of J.M. Lynch and S.H.T. Harper, ARC Research Station, Letcombe.)

the acetic acid toxicity, is fungicidal and supplies some oxygen. This example illustrates the complexity of the interactions between plants and micro-organisms and also how a change in agricultural practice, made essential or at least desirable by unrelated causes, may have unexpected side effects involving the whole complex balance between soil and plant inhabiting micro-organisms.

Conclusions

The discussions up to now have covered the microbiology of flowers leading to fruit and seed producton and finally germination and the new plant. We must now consider in the following chapters the microbiology of the leaves, stems and roots which arise from those seeds.

Flowers are too transient to have a large number of microbes, but fruits and seeds are stores of nutrients which inevitably have microbes living on them as opportunistic transients or more specialized organisms which can seriously affect the plant and destroy these food reserves. Seeds, while not themselves being affected, may carry plant pathogens to the next generation of hosts. Man, by concentrating on fruit production, storage and marketing has provided favourable conditions for many micro-organisms which cause rots of ripe or overmature fruits. Finally there may be problems in germinating seeds in unnatural conditions in nurseries which mass produce plants; the loss of many seeds in nature may be allowed for by gross overproduction by most plants. The destruction of seed which has been specially grown, tested and bought may represent a large financial loss, especially if it has wasted heated greenhouse space while it was being killed.

Selected references and further reading

Baker, K.F. (1972). Seed pathology. In Kozlowski, T.T. (Ed.) *Seed Biology*, Vol. 2. Academic Press, New York. p. 318–416.

Brown, M.E. (1974). Seed and root bacterization. *Annual Review of Phytopathology*, **12**, 181–97.

Byrde, R.J.W. & Willetts, H.J. (1977). *The Brown Rot Fungi of Fruit; their biology and control*. Pergammon Press, Oxford. pp. 171.

Campbell, R. (1983). *Microbial Ecology*, 2nd edition. Basic Microbiology no. 5. Blackwell Scientific Publications, Oxford. pp. 191.

Capriotti, A. (1953). The yeasts in flowers. *Revista Biologia*, **45**, 367–74.

Christensen, C.M. (1972). Microflora and seed deterioration. In Roberts, E.H. (Ed.) *Viability of Seeds*. Chapman & Hall, London. p. 59–93.

Davenport, R.R. (1976). Distribution of yeasts and yeast-like organisms on aerial surfaces of developing apples and grapes. In Dickinson, C.H. & Preece, T.F. (Eds) *Microbiology of Aerial Plant Surfaces*. Academic Press, London, p. 325–59.

Dennis, C. (1976). The microflora of the surfaces of soft fruits. In Dickinson, C.H. & Preece, T.F. (Eds) *Microbiology of Aerial Plant Surfaces*. Academic Press, London. p. 419–31.

Goepfert, J.M. (1980). Vegetable, fruits, nuts and their products. In International Commision on Microbiological Specifications for Foods (Eds) *Microbial Ecology of Food*, Vol. 2. Academic Press, London. p. 606–42.

Harvey, J.M. (1978). Reduction in losses in fresh market fruits and vegetables. *Annual Review of Phytopathology*, **16**, 321–41.

Lynch, J.M., Harper, S.H.T. & Pryn, S.J. (1977). Interactions between micro-organisms and barley seeds. *ARC Letcombe Laboratory Annual Report 1976*. p. 60–61.

Mirocha, C.J. & Christensen, C.M. (1974). Fungus metabolites toxic to animals. *Annual Review of Phytopathology*, **12**, 303–30.

Noble, M. & Richardson, M.J. (1968). An annotated list of seed-borne diseases. *Phytopathological Papers* 8, 2nd edition. Commonwealth Mycological Institute, Kew. pp. 191.

Pelhâte, J. (1968). Inventaire de la microflora des blês de conservation. *Bulletin Societé Mycologie de France*, **84**, 127–43.

Pepper, E.H. & Kiesling, R.L. (1963). A list of bacteria, fungi, yeasts, nematodes and viruses occurring on and within barley kernels. *Proceedings of the Association of Official Seed Analysts of North America*, **53**, 199–208.

Peterson, H.L. & Loynachan, J.E. (1981). The significance and application of *Rhizobium* in agriculture. *International Review of Cytology*, Supplement **13**, 311–31

Purchase, I.F.H. (Ed.) (1974). *Mycotoxins*. Elsevier, Amsterdam. pp. 443.

Raymond, F.L., Etchells, J.L., Bell, T.A. & Masley, P.M. (1959). Filamentous fungi from blossoms, ovaries and fruit of pickling cucumbers. *Mycologia*, **51**, 492–511.

Samish, Z., Ettinger-Tulczinska, R. & Bick, M. (1963). The microflora within the tissue of fruit and vegetables. *Journal of Food Science*, **28**, 259–66.

4
Microbiology of Living Leaves

Introduction

Leaves constitute the major visible surface of plants and are open to infection or saprotrophic colonization by air-borne or splash dispersed micro-organisms. Leaves vary greatly in their characteristics, especially with regard to microbial habitats: from the soft young leaves of a newly-opened bud to the very hard leaves of a xerophyte protected by a thick cuticle and probably extensive wax development. It is necessary to briefly consider the leaf structure before we go any further (Preece & Dickinson, 1971; Dickinson & Preece, 1976; Juniper & Jeffree, 1983).

The walls of epidermal cells are covered with a cuticle composed mostly of cutin which is a polymer of fatty and hydroxyfatty acids. The cuticle makes the leaf surface almost waterproof, especially if it is well developed, and this is further enhanced by a surface coating of wax (Fig. 4.1 A and D) which often occurs in tubules, rods or sheets, the shape being characteristic of the particular plant, though influenced by the environment. The wax is composed of alkanes, alkyl esters and aldehydes, primary and secondary alcohols and long chain monobasic organic acids. Conversely wax affects the wettability of leaves which is most important in the efficiency of sprays against microbial pathogens.

The topography of the leaf surface also has a great effect on micro-organisms; the adaxial and abaxial surfaces (usually the top and the bottom respectively) may have a different microtopography. Usually the walls of the epidermal cells give a series of troughs across the surface (Fig. 4.1 B) and veins may give very considerable raised areas on the undersurface. Most leaves also have hairs or spines which may be multi- or unicellular and may be glandular (Fig. 4.1 C). Other nooks and crannies may be formed by such structures as epistomatal cavities (Fig. 4.1 D).

The microclimate on the leaf is different from the surrounding air. The temperature is usually higher during the day because of the radiant energy from sunlight, but radiation may cool the leaf at night. There may also be quite high levels of ultra-violet light. Many leaves are also rather dry habitats, without free water for much of the time. Exceptions are the plants of very wet regions such as rain forests or cloud forests which have different leaf surface communities from plants living in drier conditions. Wind speed at the leaf surface is zero but it rapidly increases away from the leaf, the depth of the boundary layer depending on the roughness of the surface, the number of hairs and the rate of bulk air movement. Nutrients are usually in short supply, they do however arrive in dust and dirt blown from the soil (Fig. 4.1 C), and in

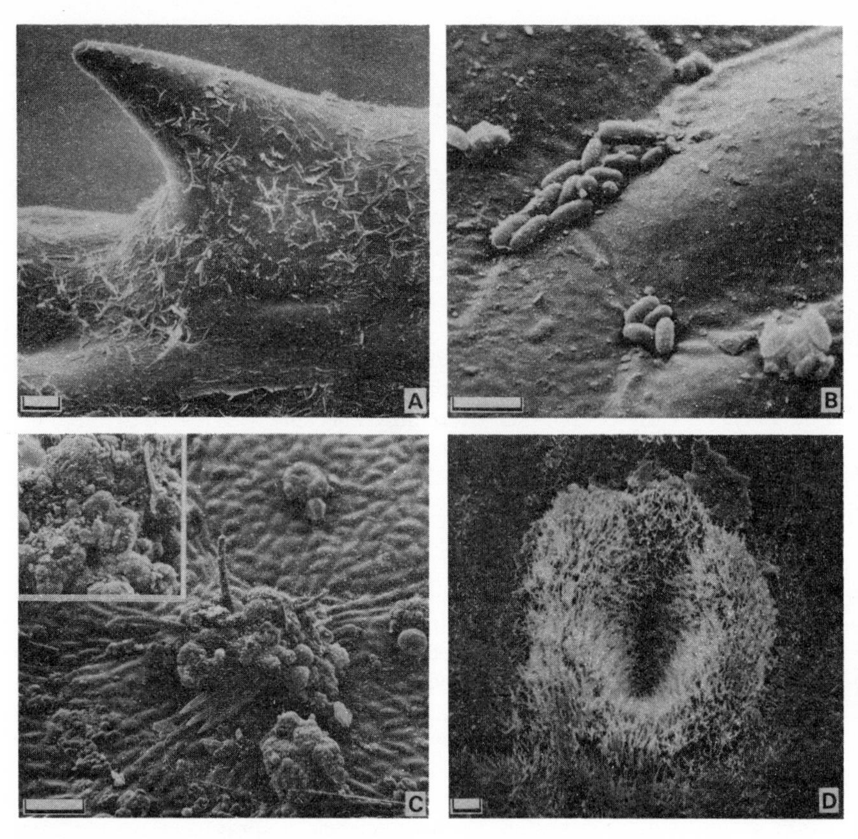

Fig. 4.1 **A.** Wax on the surface of a spine on the needle of Corsican pine. Bar = 5 μm. (From Campbell, R., 1972, *Annals of Botany,* **36**, 307–14). **B.** Depressions between epidermal cells on the surface of Hawthorn (*Crataegus*) with a few yeasts. Bar = 5 μm. **C.** Leaf hair of oak (*Quercus* sp.) with attached debris. Bar = 50 μm. Inset, detail to show spores of *Cladosporium* on the debris. **D.** Epistomatal cavity on the needle of Corsican pine, filled with tubular wax. Bar = 5 μm. (From Campbell, R., 1972, see A.)

pollen (p. 65). Many substances, some of which may be microbial nutrients, are leached out of leaves. This is again affected by the type of plant, the leaf age, etc., which determine the cuticle and wax characteristics and hence the permeability of the leaf. Inorganic ions (K^+, Ca^{2+}, Mg^{2+}, Mn^{2+}) are leached or conversely may be taken up from incident rain water or drip from higher layers in the canopy. Epiphytes, especially in tropical and temperate rain forest, may greatly alter the nutrients dripping down through the canopy: for example lichens with cyanobacterial symbionts may make significant contributions to the available nitrogen (p. 84), but other epiphytes remove nutrients. Many organic materials, especially carbohydrates and organic and amino acids, are also leached from leaves by rain and dew. There are few quantitative data but the amount per unit area of leaf is not large unless there is aphid or scale insect infestation when the damage to leaves and the production of 'honeydew' may allow large quantities of nutrients to be released. Under these conditions a very obvious microbial flora is produced, comprised mainly of yeasts and the so-called sooty moulds, which may blacken the leaves. Apart from the loss of carbohydrate the dark fungi may reduce photosynthesis by shading.

Finally in the consideration of nutrients, guttation from hydathodes may also contribute amino acids, carbohydrates and minerals and the guttation fluid frequently contains micro-organisms (*Pseudomonas fluorescens*, *Bacillus cereus*, various yeasts and *Cladosporium*). Infection by some pathogens may occur through the hydathode after growth in the guttation fluid (e.g. *Xanthomonas campestris*).

Micro-organisms grow in micro-habitats which ameliorate some of these conditions on the leaf surface (Fig. 4.1 B and C). A depression in the leaf surface, an epistomatal cavity or a spine can give some protection from sunlight and perhaps higher humidity. Water will also drain down into the gutters between epidermal cells.

The leaf surface may also contain substances which inhibit microbial growth. These include general water-soluble exudates such as gallic acid in Norway maple, diterpenes from glands on tobacco or compounds that specifically inhibit pathogens but not saprotrophs.

Both saprotrophs and pathogens grow on and in leaves (epiphytes and endophytes). They arrive from air-borne propagules or soil and dust particles and after growth they may release spores into the air again. Phyllosphere organisms (those growing on or near the leaf) are of importance because leaf diseases in agricultural and forest crops can have a dramatic effect on the yields or can kill the plants. It is necessary to know about the pathogens, and the potentially antagonistic saprotrophs, to control diseases and to understand the effects of pesticide applications. Recent studies have also shown the importance of the leaf-surface organisms in some natural ecosystems in terms of their biomass and activities in cycling, producing or immobilizing minerals and organic nutrients. The decay of leaf litter is considered in Chapter 7 (p. 153).

Methods of study

The main methods of study are washing and plating, and direct examination (p. 13). Dilution plates have all the problems previously discussed and it is usually rather difficult to dislodge a reasonable proportion of the microbes in the washing: glass beads and detergents have been used to assist removal.

Direct examination with the light microscope requires that the leaves be made transparent by decolourizing before staining. Such methods are obviously destructive and do not permit resampling. Micro-organisms can be removed by pressing transparent sticky tape onto the leaves, then staining and observing the tape microscopically. Nail varnish or other cellulose lacquers can be painted onto leaves, allowed to dry and then peeled off together with most of the micro-organisms. There have been a few studies to estimate biomass by measuring the cross-sectional area of microbes on serial leaf sections: these yield valuable information on the distribution of the biomass but are too laborious to be widely used. Epifluorescence using ultra-violet light has been used most successfully on leaves. Methods to measure microbial activity in terms of respiration, etc., have great difficulty in separating microbial and leaf metabolism.

In general then, routine microbiological methods have been used with leaf-surface microbes and have given information on their numbers (largely of unknown taxa), or there is rather better identification but more dubious numerical data from culture work. Most studies use direct examination in some form as the quickest way to get an estimate of numbers or biomass and some indication of the distribution patterns. The sparseness of colonization may prove a problem; at high enough magnifications to see the organisms the fields of view may be so small that many will not contain any microbes.

The methods used for leaf surface examination have been extensively reviewed by Preece & Dickinson (1971), Dickinson & Preece (1976), Blakeman (1981) and Dickinson (1982).

Bud microflora

The flora of the interior of the bud is different to that of the external bud scales which are exposed to the air. The leaf primordia are frequently sterile or carry very small microbial populations. In particular, actively-growing filamentous fungi are usually rare and most of the organisms are apparently dormant, using the bud as a place in which to survive unfavourable conditions before colonizing the leaves when the bud opens and the leaves expand. The filamentous fungi that are isolated (Fig. 4.2) are usually heavily sporing '*r*' strategists such as *Coniothyrium, Alternaria, Phoma* and *Cladosporium* and there is no evidence that they are other than dormant spores.

The bud microflora is dominated by yeasts (up to 10^6 g^{-1} wet weight, Fig. 4.2), especially white forms but also including pink yeasts and *Aureobasidium pullulans*. Some of these may be actually within the leaf primordia rather than on their surface. Fungal pathogens also occur within buds. *Taphrina*, the cause of various leaf curl diseases (p. 68), overwinters on shoots or under bark, but infects buds just before opening: it is important to apply fungicide before the organisms are protected in the bud. *Tilletia caries* (stinking smut) is systemic in the plant, including the apical meristems. Another smut (*Ustilago maydis*) occurs in axillary buds of maize. Various pathogens have been shown to overwinter in the protected environment of the bud: for example two apple diseases (scab, *Venturia inequalis* and mildew, *Podosphaera leucotricha*) are present in buds as dormant mycelia.

Bacteria are also present in numbers from 10^5 to 10^6 g^{-1} wet weight (Fig. 4.2. There is very little information as to what these might be, but fluorescent pseudomonads are apparently quite important (Fig. 4.2). Some bacterial pathogens are also known to occur in buds. *Erwinia amylovora*, the cause of fire blight in fruit trees, *Pseudomonas syringae*, and *P. glycinea* in soybeans are reported from apparently healthy buds. There are also some specialized cases in which symbiotic bacteria (probably mutualistic symbionts) live in buds. Plants of the Rubiaceae and Myrsinaceae frequently have leaf nodules which contain bacteria when young. The bacteria (*Chromobacterium, Mycobacterium, Klebsiella* and *Xanthomonas*) probably provide the plant with cytokinins or other hormones; they occur in the flowers, are passed to the seed

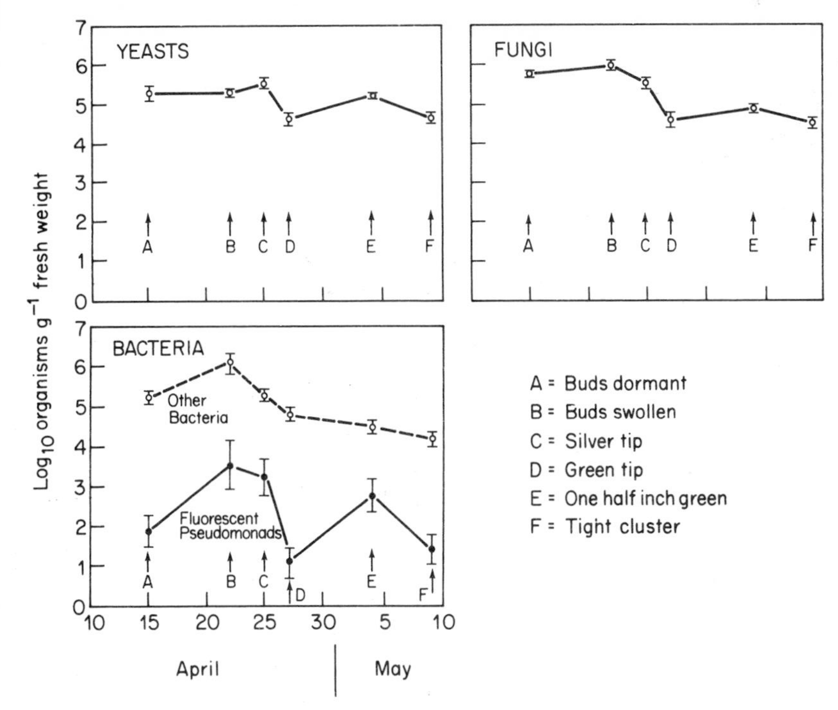

Fig. 4.2 Population densities of yeast, filamentous fungi and bacteria from apple buds. Bud expansion is indicated by the letters A → F. (From Andrews, J.H. & Kenerley C.M., 1980, *Canadian Journal of Botany*, **58**, 847–55.)

and infect the new plant. They are in high numbers in the mucilage within buds from where they infect the developing leaves.

As far as is known there are no reports of protozoa or algae in buds, though viruses do occur. The latter may stimulate the axillary bud development and result in witches broom symptoms or the bud may be enlarged, or die, or produce short yellow shoots which die later in various 'yellows' diseases. Some of these 'virus' infections of buds which have been carefully investigated, e.g. tomato big bud, some yellows and witches brooms, have been shown to be caused by mycoplasmas (p. 5).

As the bud begins to open the numbers of oganisms per unit weight tends to decrease (Fig. 4.2). This is however on a wet weight basis and may reflect water content changes. On a unit area basis the numbers decrease greatly because the leaves expand much faster than the microbes can multiply to cover the surface.

Saprotrophic microflora of leaves

Temperate deciduous plants

Leaves of temperate agricultural crops and angiosperm (broad-leaved) trees are comparatively short-lived. They have been quite well studied and have a basically similar microflora regardless of species, though some pathogens are specific. Conifers, with needles living as long as 8 years, give rather different conditions and tropical vegetation, especially if evergreen, is different again and these two will be considered later (p. 59 and 63).

There seems to be a remarkably similar saprotrophic flora on all temperate, deciduous annual plants. The filamentous fungi are mainly *Cladosporium* spp. (*C. herbarum* or *C. cladosporioides* especially), and *Alternaria* and *Epicoccum*. The yeasts are dominated by *Aureobasidium pullulans*; pink yeasts (e.g. *Sporobolomyces*) and the white yeasts such as *Cryptococcus* are common; *Candida humicola*, *Torulopsis* and *Trichosporon* are frequently reported. Bacteria (Blakeman, 1982) from leaves are often chromogenic: pseudomonads are the most common (especially *P. fluorescens*), and *Chromobacterium*, *Klebsiella*, coryneforms and *Bacillus* are usually present. *Flavobacterium* and *Erwinia herbicola* seem to be rather variable, being reported as common by some authors but not by others. There are very few anaerobes or actinomycetes on leaf surfaces.

It should be stressed that these lists are the results of cultural studies and may be biased by the media provided. They may also not distinguish between actively-growing forms and those present as resting structures. For example though *Aureobasidium pullulans* is ubiquitous there is good evidence that it forms chlamydospores and is dormant for the later summer months, even though it is routinely isolated. Those organisms which do not form obvious resting structures may still not be active because of environmental constraints; humidity and water availability have been especially studied. In general high humidity or free water are essential for growth. The free water may be as rain, but dew is very important in some cases since it may be present for many hours each day at some times of the year, or in some habitats, even when the weather is usually dry. In experimental studies *Alternaria* spores did not germinate, or produced only very short germ tubes, at 70 to 80% relative humidity and the spores lost viability; only 11% of *Cladosporium* spores germinated at this humidity. Both fungi germinated and grew at 97% relative humidity. With *Sporobolomyces roseus* a relative humidity greater than 85% or the presence of dew was necessary for the growth of this yeast after artificial inoculation (Fig. 4.3).

These environmental effects may well be the main cause of the differences in microbial numbers or species which occur on different parts of leaves or in different positions in the canopy. Numerous studies have shown that lower surfaces (usually abaxial) of leaves have higher microbial populations than the upper. Lower surfaces are shaded and are also protected from some of the effects of surface wash-off. There are also reports of other positional effects. Andrews *et al.* (1980) showed that in apple trees the height in the canopy and whether the leaf was on the outside of the tree or amongst the branches in the middle of the crown had an effect on yeast, filamentous fungi and bacterial populations, and the latter were also affected by the orientation to north, south, east or west. This was explained in terms of the availability of air spora for inoculation, outside leaves having the advantage, and the microclimate inside the canopy which was characterized by higher humidity and longer periods of wetness. The study was, because of the number of measurements made, carried out on rather a small number of trees but it points out some of the problems to be borne in mind when trying to compare different plants, or

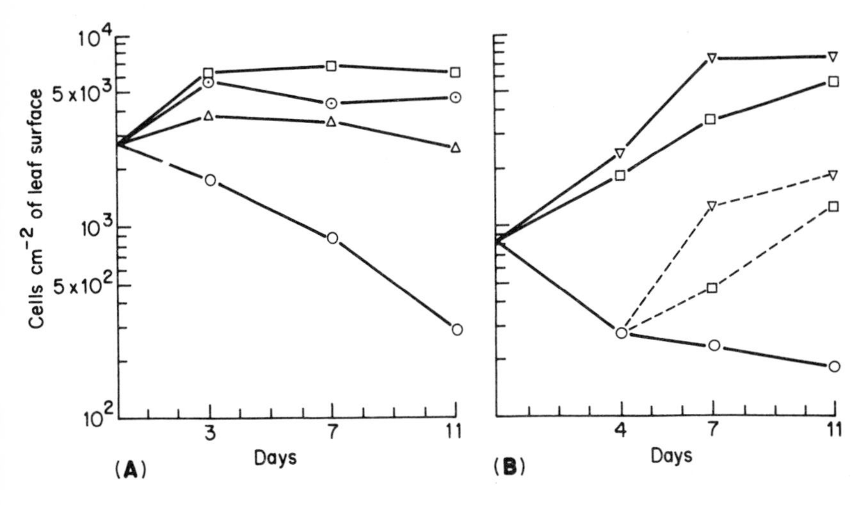

Fig. 4.3 **A.** Effect of different relative humidities on the development of *Sporobolomyces roseus*. 65% RH (○——○); 75% RH (△——△); 85% RH (⊙——⊙); 95% RH (□——□). **B.** Effects of dew (8 h dew and 16 h RH 95%), high and low RH on development of *S. roseus*. Dew (▽——▽): 95% RH (□——□): 65% RH (○——○): 65% RH to dew (○ — — ▽): 65% RH to 95% RH (○ — — □). (From Bashi, E. & Fokkema, N.J., 1977, *Transactions of the British Mycological Society*, **68**, 17–25.)

worse, different species or vegetation types. Sampling must be done to try to ensure that these positional effects are eliminated to give genuine comparisons between plant types and not, for example, between a north-facing, outside canopy at 5 m and a south-facing, within canopy sample at ground level. It may be possible to show differences in the micro-organisms on leaves of species X and Y, but whether it is due to some intrinsic characteristic of X and Y or to some other variable environmental factor may be difficult to say.

There are also seasonal effects on the microbes on leaves. The numbers of all organisms are low at the opening of leaves (p. 55). The population soon increases however as the bud flora multiplies and new inoculum arrives from the surrounding air. Numbers usually reach a peak in autumn as the leaves senesce. These general trends in the numbers of yeasts, filamentous fungi and bacteria are shown in Fig. 4.4; these data come from a study of apple trees and the effects of pesticides to which we will return later (p. 73), at present look only at the control (dashed lines). The data are from dilution plates so the spore-forming fungi are favoured and this may not reflect mycelial growth. In another study on *Acer* the mycelial growth increased from July to October and was mainly of *Cladosporium*.

Even though the numbers of organisms are quite large (10^5 or 10^6 g^{-1} wet weight) the proportion of the leaf surface covered by micro-organisms is small. Fungi generally cover less than 5% of the leaf surface and bacteria, because of their small size, may range from 0.0001% to a maximum of 0.1% cover.

These general figures mask the specific differences in colonization patterns (Fig. 4.5). Some of the fungi such as 'sterile dark 2' and *Acremonium charticola* occur in the buds and colonize young leaves but do not persist.

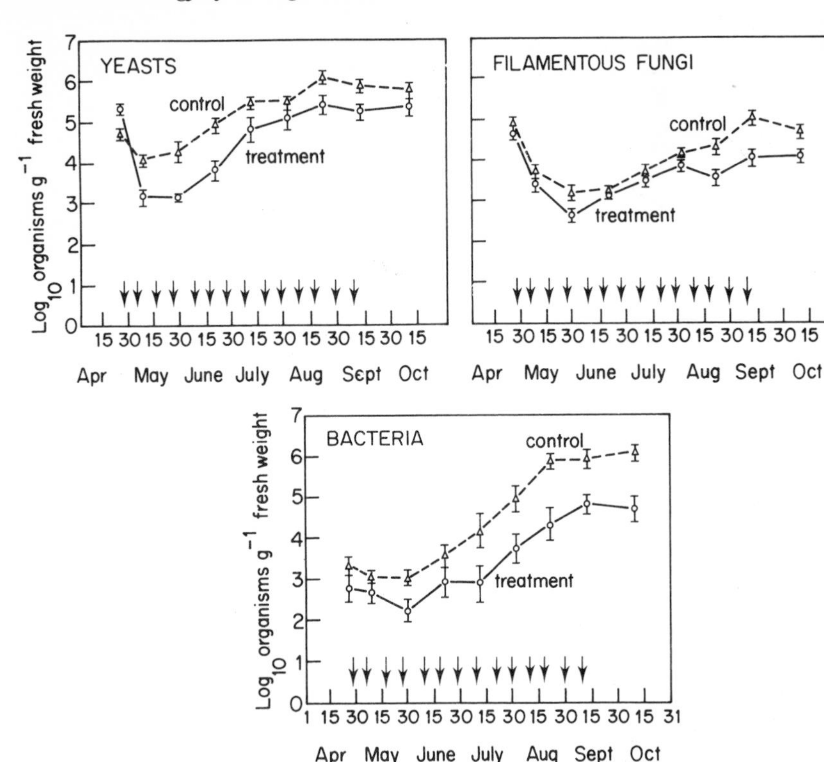

Fig. 4.4 The numbers ($\log_{10}$ g^{-1} fresh weight) of yeasts, filamentous fungi and bacteria on apple leaves. Control leaves (Δ — — Δ) were untreated but others (O———O) were sprayed (at ↓) with fungicide and insecticide. (From Andrews, J.H. & Kenerley, C.M., 1978, *Canadian Journal of Microbiology*, **24**, 1058–72.)

Aureobasidium pullulans however is in the bud, colonizes and grows to maximum numbers in later summer and persists on the fallen leaves. White yeasts colonize from the air in summer but do not persist after senescence. *Cladosporium* sp. colonize early and persist in high numbers but *Alternaria* and 'sterile dark 1' are mainly found on senescent and fallen leaves. It is possible, indeed probable, that similar differences occur amongst the bacteria, but there are no data, largely because of the problems of identification.

The rise in numbers of saprotrophs at senescence may be important agriculturally, especially in cereals where a large proportion of the grain weight is dependent on the photosynthesis in the upper leaves. If photosynthesis can be prolonged by delaying senescence then the yield is increased. Senescence is associated with this rapid rise in microbial numbers, but are the microbes causing senescence or are they merely responding to the increased exudation from dead and dying leaves? The obvious experiment of applying wide spectrum fungicides, e.g. benomyl, has been done and there are reductions in the fungi: there is also a persistence of chlorophyll and the yield increases. It is still disputed whether the microbes cause the senescence because, certainly with benomyl, there are cytokinin-like effects which increase chlorophyll retention

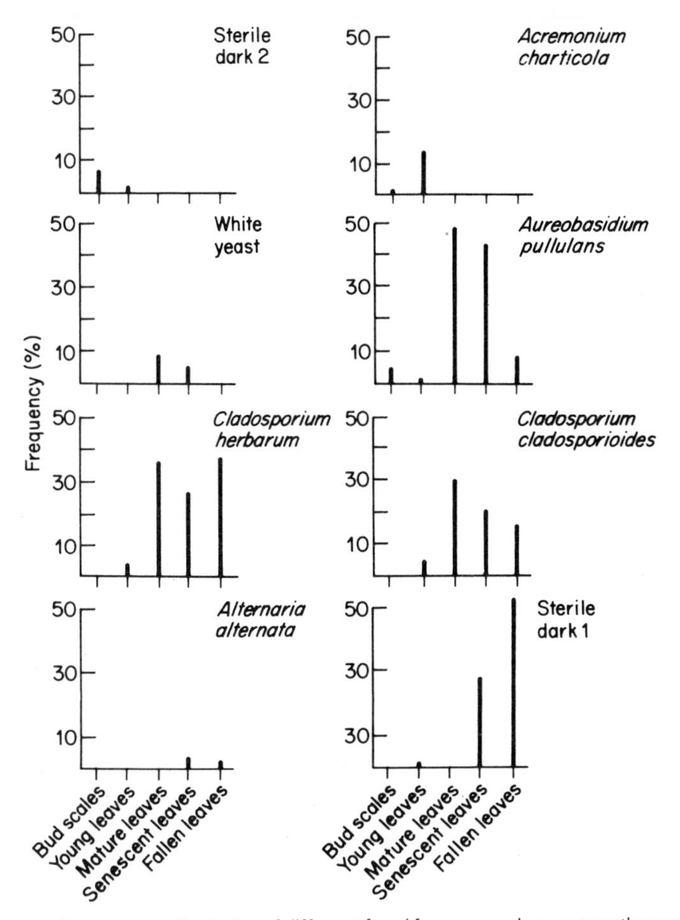

Fig. 4.5 Frequency of isolation of different fungi from aspen leaves over the season to show different patterns of colonization and variation in numbers of different organisms over the season. (Selected from Wildman, H.G. & Parkinson, D., 1979, *Canadian Journal of Botany*, **57**, 2800–11.)

regardless of the reduction or otherwise in the numbers of microbes. Even though other fungicides also give yield increases, the mechanism of action may not be entirely by affecting the microbes. There are other advantages in later applications, especially that the grain is 'cleaner' and less infected with surface moulds (p. 43) which increases the quality and hence the price, regardless of yield effects.

Microflora of conifer needles

There have been comparatively few studies on the microflora of evergreen conifers. Those that have been done suggest that the organisms involved are much the same as for the deciduous annual leaves, including the yeasts *Sporobolomyces*, *Rhodotorula* and *Aureobasidium*, with filamentous fungi

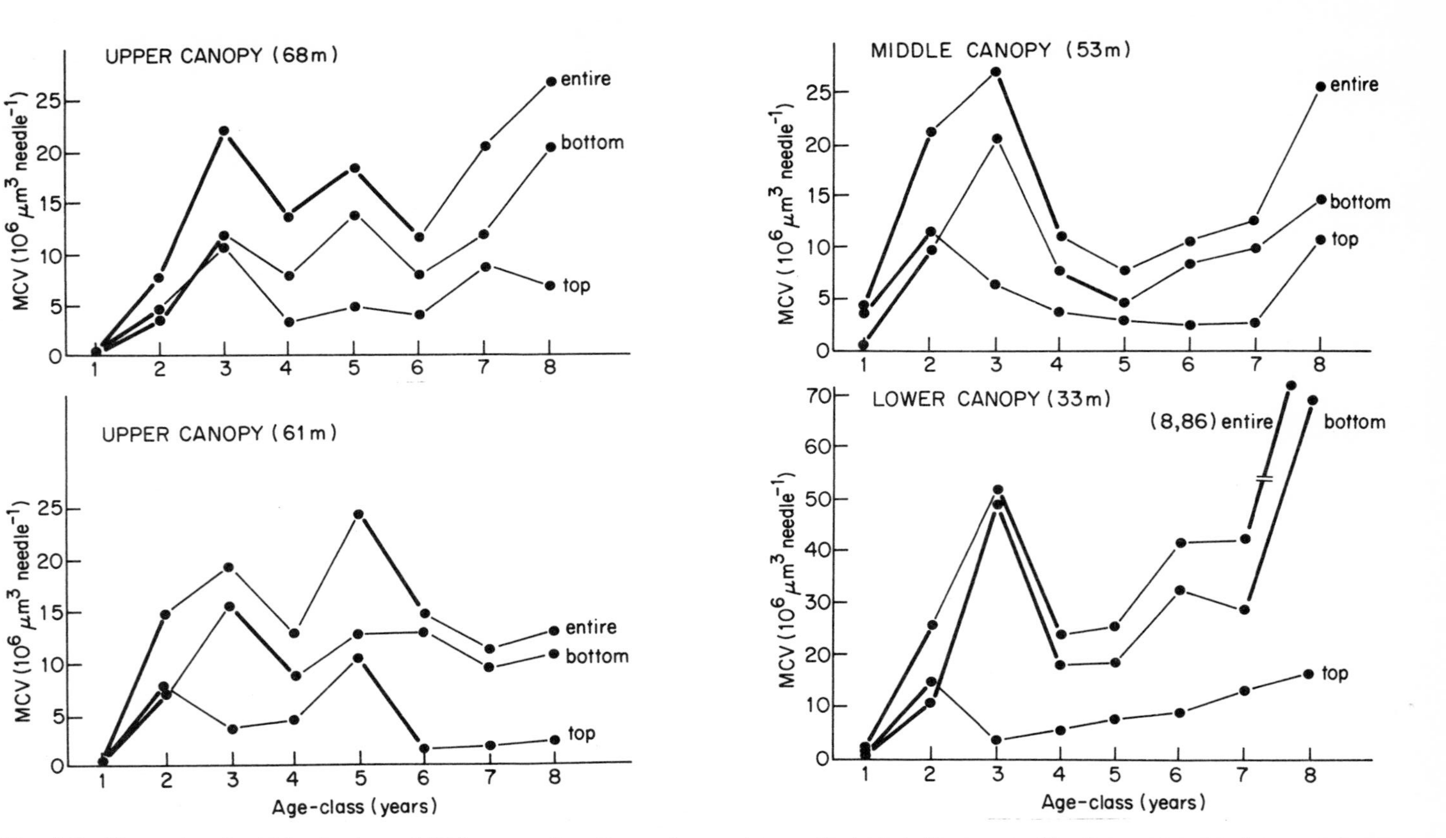

Fig. 4.6 Change in microbial cell volume (MCV) per needle with needle age at several heights in the canopy. Needles produced in the current year, beginning with bud burst in June, are considered to belong to age-class 1. Thicker lines indicate differences in estimates of population means for adjacent age-classes are significant at P = 0.1 level. (From Carroll, G.C., 1979, *Canadian Journal of Botany*, **57**, 1000–7.)

such as *Cladosporium* and *Epicoccum*. The common leaf-surface bacteria (*Pseudomonas, Flavobacterium, Xanthomonas*) are also present. These are '*r*' strategist microbes; the main difference is in the persistence of the needles which allows slow-growing '*K*' strategist epiphytes, even lichens, to become important and endophytes may become more common in later years.

Most of the recent work has been done by G.C. Carroll (e.g. Figs 4.6 & 4.7; Table 4.1) and his associates in the wet conditions of the western slopes of the mountains of north-west U.S.A. and this may not give data representative of some of the drier conifer sites. As in deciduous leaves the microbial growth is concentrated in the grooves on the leaves and in other protected micro-habitats such as the epistomatal cavities (Fig. 4.1 D). Such zones occupy about 35% of the conifer needle surface but have 70% of the microbial cell volume.

The top surfaces of the needles are colonized faster than the lower surfaces and so have a greater microbial volume for the first two years, but the top is more exposed to ultra-violet light and has less protected micro-habitats so that by their third year there is more microbial volume on the lower surface (Fig. 4.6). There are also differences in the biomass distribution throughout the canopy, regardless of the needle age or surface. Lower canopy levels have a higher biomass possibly because of the greater shelter and the extra nutrients that are leached from the needles above them. When the needle surface area is considered, year 1 is very sparsely colonized and year 3 needles have the maximum microbial cell volume: over 40% of the cell volume is on third year needles. The total standing crop of epiphytes on Douglas fir is probably 38–60 kg ha^{-1} which is small compared with the standing crop of the trees or of microbes in soil or litter. Microbial biomass turns over 2 to 5 times per year.

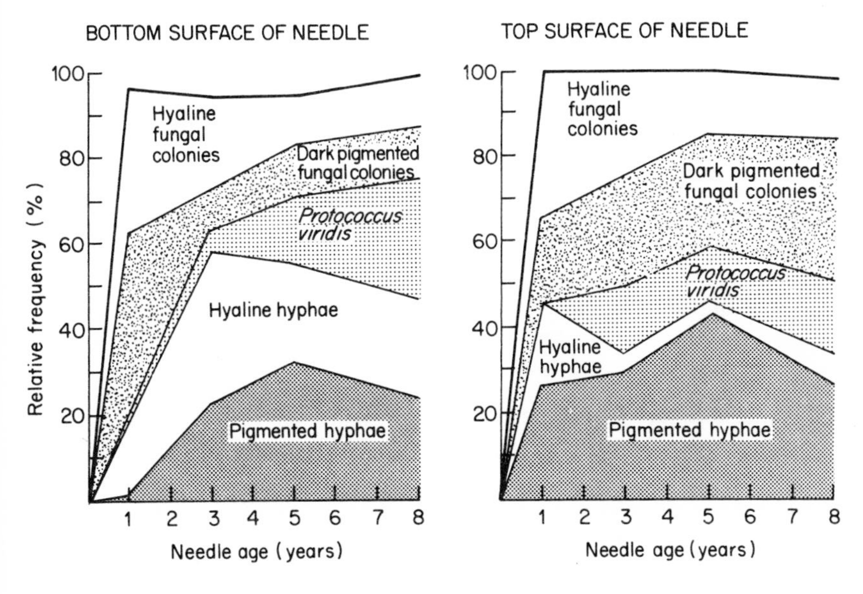

Fig. 4.7 The frequency of isolation of different groups of micro-organisms from the top and bottom surfaces of conifer needles. (Based on data of Bernstein, M.E. & Carroll, G.C., 1977, *Microbial Ecology*, **4**, 41–52.)

These biomass figures are rare in the literature and are from an intensive study of very few trees. Just how this biomass is made up is illustrated by further work of Carroll's group on the temperate coniferous rain forest (Fig. 4.7). The wet climate is reflected in the importance of algae, especially *Protococcus*, which is present on all needles. Hyaline hyphae are more common on the lower surface of the needle and pigmented hyphae and pigmented colonies are dominant on the upper surface: this distribution may reflect the tolerance to ultra-violet light which is needed on the upper surfaces.

The increase in the number or biomass of micro-organisms at senescence may not be the maximum on conifers as it is on annual leaves (compare Fig. 4.7 with Fig. 4.4). However the overall biomass and percent of the needle covered is high compared with annual leaves. This may reflect the longer time available for colonization and the ability of some slower-growing organisms to become estabilshed.

Some of the needles (and more or less all of the small branches, p. 84) have crustose lichens. The commonest one in Carroll's study was *Lobaria oregana*. Lichens containing cyanobacteria may make a significant contribution to the nitrogen economy of Douglas fir canopies. The nitrogen is transferred to bacteria (*Pseudomonas* and *Arthrobacter*) which grow in very high numbers on the lichen surface (10^5 g^{-1}, or even 10^7 in wet periods). These bacteria are washed off, down through the lower canopy layers and they take the fixed nitrogen with them. Lichens and their associated bacteria are the basis of many canopy food chains.

There is very little information about the bacteria on the needles themselves, but numbers are probably low (10^4 g^{-1}) and this is consistent with their rather infrequent appearance when examining conifer needles with the scanning electron microscope.

The occurence of apparently saprotrophic fungi within leaves has been mentioned (p. 54) and they seem to be widespread in conifers, probably again reflecting the longer life of the needles. Most needles have multiple infection in the first year and the infection rate increases in Douglas fir up to 3 year-old needles. In juniper, Lawson's cypress and western red cedar infection rates are lower, though again they increase with needle age reaching a maximum of 30 to 50% of needles infected. *Taxus*, *Sequoia*, sitka spruce, *Abies magnifica* and *A. grandis* have about 60 to 100% of the needles infected (Table 4.1). The frequency of infection is greatest at lowland, wetter sites. In some hosts endophytic fungi may be more common in the petioles rather than the lamina (*Picea breweriana* and *P. englemanii*) and vice versa (*Pinus attenuata*) (Table 4.1). The fungi involved generally have a broad host range though a few seem to be host species specific. The common endophytes are sometimes also epiphytes (*Aureobasidium pullulans*, *Cladosporium*, *Alternaria*); there are some ascomycetes and many pycnidial-forming deuteromycetes which occur as endophytes. *Phyllosticta concentrica* is especially common together with *Phomopsis*, *Cryptocline* and *Leptostroma*.

Most of these endophytes are saprotrophic and only rarely produce obvious lesions or fruiting bodies on the needles. A few of them are however definitely latent infections by pathogens. *Leptostroma pinastri* in *Pinus sylvestris* is the imperfect stage of *Lophodermium pinastri* which causes needle cast in older

Table 4.1 Endophytic fungi in confer needles. (Selected from Carroll, G.C. & Carroll, F.E., 1978, *Canadian Journal of Botany*, **56**, 3034–43.)

Host	% infected needles (all fungi)	% petiole segments infected with petiole fungi	% petiole segments infected with blade fungi	% blade segments infected with blade fungi
Abies amabilis	20.4	2.4	5.4	7.2
Abies concolor	49.5	1.0	32.7	36.2
Abies grandis	66.6	2.4	52.1	50.9
Abies lasiocarpa	21.1	0.0	8.9	9.3
Abies magnifica	85.1	2.4	62.1	54.8
Abies procera	42.3	0.0	17.6	26.1
Picea breweriana	36.3	34.0	0.0	1.6
Picea engelmannii	56.4	52.7	0.9	3.3
Picea sitchensis	94.8	34.9	77.2	68.5
Pinus attenuata	70.0	0.0	0.0	18.6
Pinus contorta	46.0	2.7	14.4	14.6
Pinus lambertiana	67.6	0.0	32.4	33.6
Pinus monticola	92.3	8.1	28.8	44.0
Pinus ponderosa	77.4	14.2	23.5	21.6
Pseudotsuga menziesii	71.3	9.6	58.0	54.0
Sequoia sempervirens	100.0	0.0	82.4	97.0
Taxus brevifolia	24.6	7.3	9.6	10.2
Tsuga heterophylla	39.0	20.8	9.8	8.2
Tsuga mertensiana	56.1	19.0	9.4	19.0

needles. The development of the lesions and the pathogenic phase is usually dependent on some injury or stress such as drought to initiate symptom expression.

The microflora of conifer needles is obviously different from that of annual angiosperms and the rather limited data (almost all from north-west U.S.A.) suggest that this is related to the longevity of the needles as well as basic anatomical and chemical differences between the two types of vegetation.

Microflora of tropical leaves

Again there is very little information on these leaves, though like the conifers such forests cover a large area of land surface and are very significant in world biomass terms. The 'tropics' vary enormously in climate, altitude, etc. Nothing is known of the leaf surface flora of arid or desert zones. Tropical rain forest may have many perennial leaves and the humidity and rainfall are high. The major work has been done by Ruinen (1961) who studied mostly Indonesian forests, and there are many reports of particular fungi being described from leaves, usually in the course of taxonomic studies. Such studies rarely give information on the importance of the organism in the leaf surface community.

In tropical rain forest conditions the phyllosphere is very different to that just described for temperate crops. Often there is a continuous microbial layer up to 20 μm thick on mature leaves. The first colonists are bacteria, especially nitrogen-fixing organisms such as *Beijerinckia*, which is pan-tropical.

Cyanobacteria, actinomycetes and fungi form a network of filaments with the spaces filled with bacteria. Algae, yeasts and lichens then colonize the microbial layer followed by flagellates, myxomycetes and ciliates. Typically there is free water at all times in this microbial layer. The biomass may be very high and figures of up to 4387 kg dry weight ha^{-1} have been calculated.

Different tree species seem to have different floras. For example, a study on three tree species in Hawaii (*Metrosideros collina* var. *polymorpha, Acacia koa* var. *hawaiiensis* and *Cheirodendron trigynum* var. *trigynum*) showed that they had very different, though extensive, fungal floras. There were 35 genera of cosmopolitan fungi recorded, 90% imperfect fungi. By far the majority of these were on *Metrosideros* whose leaves had 118 taxa on them, 69 of which were characteristic of this particular host (Table 4.2). It must be borne in mind that this host has a wide climatic and altitude range and the concern expressed above (p. 57) about whether intrinsic differences are being recorded or merely differences in microclimate applies in this case.

Table 4.2 Occurrence of fungi on leaf surfaces of three endemic vascular plants in Hawaii. (Modified from Baker, G.E. & Dunn, P.H., 1979, *Mycologia,* **71**, 271–92.)

	Total number	Only on *Metrosideros*	Only on *Acacia*	Only on *Cheirodendron*
Zygomycotina	4	2		1
Ascomycotina	9	3	2	1
Basidiomycotina	4	3	1	
Fungi Imperfecti	151	61	18	7
Sphaeropsidales	20	9	2	1
Melanconiales	10	4		
Moniliales	121	48	16	6
Moniliaceae	53	24	5	4
Dematiaceae	54	20	8	2
Stilbaceae	0			
Tuberculariaceae	12	4	2	
Mycelia Sterilia	2		1	
Total number of species	168	69	21	9

Algae in particular have been studied on tropical leaves (Round, 1981) and the filamentous *Trentepohlia*, *Cephaleuros* and *Phycopeltis* are most commonly reported. There are also algae which live in the epi- and sub-stomatal chambers of leaves, anchoring themselves by special 'haustoria-like bodies'.

Ruinen reported the bacteria on leaves to be *Beijerinckia*, *Azotobacter*, *Pseudomonas*, *Clostridium* and *Spirillum* but the investigation was not extensive and much remains to be done. Fungi present include the common leaf-surface organisms and also the sooty moulds (Capnodiaceae). There is nothing known about the pathogens of leaves in tropical forests; there are only lists of organisms known to cause some leaf diseases of the major tropical crops.

Considering the rate at which tropical forest is being cleared, or at least profoundly altered by man's activities, we desparately need more information on the nature, numbers, biomass and activities of leaf surface microbes from these vegetation types.

Leaf pathogens

Information on leaf pathogens is again very biased towards western agricultural plants. There is some information on temperate tree diseases caused by fungi but very little on bacterial and viral pathogens of tree leaves. Tropical leaf pathogens are of course known from crop species, but there is very little understanding of their biology: control measures usually involve fungicides, etc., designed for western agriculture, that have been found to work empirically.

Pathogens usually grow at least partly within the leaf, though some such as *Erysiphe* only penetrate the epidermal cells and most of their biomass is on the surface. Pathogenic fungi can also be present in the leaves in a latent form which may subsequently develop into active infections (see above, p. 62 for conifers). *Alternaria tenuis, Botrytis cinerea* and *Fusarium* spp. have all been found within surface sterilized potato leaves that showed no visible symptoms of disease.

The pathogens of leaves are predominately fungi, which may be facultative or obligate necrotrophs and biotrophs (p. 22). The facultative necrotrophs include those organisms which colonize the senescent leaves and which may be weak, unspecialized pathogens. There are many necrotrophs, especially in the ascomycetes and imperfect fungi which cause a wide variety of leaf spots, rots and other diseases. Biotrophs include the rusts, powdery and downy mildews and of course the viruses.

The microbes have to arrive on the leaf surface, germinate if they are fungal spores, gain entry to the host, and obtain nutrients. Even obligate biotrophs have to survive and possibly grow on the leaf surface before they gain entry and there will be competition here with saprotrophs or other pathogens. It is usually at these vulnerable stages that leaf diseases can be controlled: once the pathogen is inside the leaf it is colonizing an environment for which it is specially adapted, where the competition is reduced because of the exclusion of saprotrophs and where the pathogen has available food supplies. The pathogen may however have to combat host defense mechanisms (p. 33).

Fungal pathogens usually arrive on the leaf as spores, some of which (often a low percentage) germinate and grow. This requires high humidity and/or free water and nutrients. The latter may be endogenous in some cases, the spore being large enough to carry its own reserves for the initial stages of growth. Many pathogenic fungi are however dependent on nutrients on the leaf surface which are in short supply. Pollen may serve as a nutrient source, mostly by having the nutrients leached out of the grains, and its presence may encourage necrotrophic pathogen growth or colonization. *Helminthosporium sativum* and *Septoria nodorum*, two cereal necrotrophs, had ten times as much surface growth, increased numbers of infection sites and four times the necrotic leaf area when pollen was present. *Puccinia recondita*, an obligately biotrophic rust, was not however affected by the presence of pollen. The converse of this is also true: if nutrients can be removed from the surface then infection by necrotrophs is reduced (p. 52).

The growth across the leaf surface may be random, though some hyphae do run in the grooves between the cells as previously discussed (p. 53). Some pathogens do, however, have very well orientated growth patterns, crossing over the epidermal cells at right angles. This has been shown for some rusts and for *Erysiphe* (Fig. 4.8A).

Penetration by fungi may be directly through the cuticle, usually after forming an inflated region, the appressorium (Fig. 2.3), above the point of penetration. One mycelium may penetrate many times as in the case of *Erysiphe*. Other pathogens penetrate only once and then proliferate within the leaf. This is the case with *Lophodermella*, a facultative necrotroph on pine needles. This fungus may enter through the stomata (Fig. 4.8B) as do many rusts. They penetrate the cuticle in the substomatal cavity. Exactly how penetration is achieved is not clear, a combination of enzymes dissolving the host wall and also mechanical pressure seems to be involved.

Once inside the leaf there are two basic strategies followed by fungi. Firstly there are necrotrophs whose hyphae penetrate between and into the cells killing them by penetration or by toxin production which may extend some distance in front of the advancing hyphae. Secondly there are biotrophs which though they may penetrate the cell wall rarely break the plasmalemma: the plasmalemma becomes invaginated to enclose the specialized hyphal branch, the haustorium (Fig. 2.3). Haustoria vary from species to species but the host cell remains alive and nutrients pass across the membranes to the fungus.

Penetration into leaves by pathogenic bacteria differs from that of fungi. Bacteria are dependent on openings in the leaf surface and motility seems to aid bacterial entry. The openings may be either natural ones like stomata, or natural or man-made wounds. The wounds may be very small and can be caused by wind abrasion with soil particles or by hail damage. In general, leaf pathogenic bacteria are necrotrophs and spread intercellularly through the leaf though they can, and do, spread along the lumen of xylem vessels if introduced into the cut end in wounds. Free water films are needed for invasion. One of the effects of infection is often to induce water soaking of the tissue by damage to the plasmalemmas and this favours continued invasion by the bacteria.

Apart from these, and other, physical barriers to infection and invasion there are also chemical barriers that have been discussed previously in general terms (Chapter 2 and references: Dickinson & Lucas, 1982; Heitfuss & Williams, 1976; Rhodes-Roberts & Skinner, 1982).

There are so many fungal diseases of leaves that it seems invidious to select one or the other for special mention, so a few examples will have to suffice as representatives of the thousands available which range from trivial leaf spots of apparently unimportant plants to diseases which cost millions of pounds or dollars each year in control measures, added to which there are the losses which still result during growth and marketing (see losses for lettuce in Table 3.1).

Cereal leaf diseases are of major world importance. *Puccinia striiformis*, the yellow rust of wheat and barley, is the most common cereal rust in the world and is especially prevalent in the cooler, wetter growing regions such as north-west Europe and western America. There are host specific strains. The leaves develop a yellow stripe along their length as uredia form in pustules. Photosynthesis is reduced and hence there is a reduction in vigour, in root

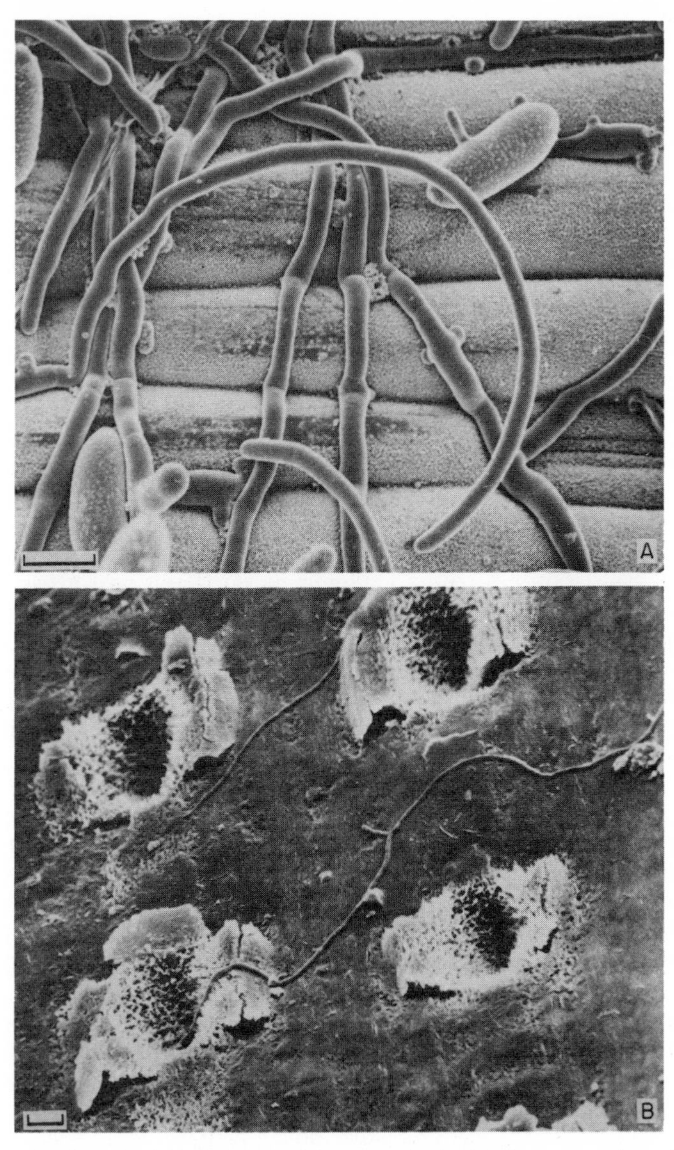

Fig. 4.8 A. Growth of *Erysiphe graminis* across the leaf surface, approximately at right angles to the length of the cells. Bar = 10 μm. (Photograph courtesy of A. Beckett, Department of Botany, University of Bristol.) **B.** Penetration of *Lophodermella* into the stomata of a pine needle. This fungus also produces appressoria and penetrates directly through the epidermis. Bar = 10 μm.

growth, and in the number of tillers which over-all results in the grain yield being much reduced. There are resistant varieties of wheat and barley and also extensive chemical control with systemic fungicides such as tridemorph, carbendazim and triadimefon.

Powdery mildews are also very important on cereals and are caused by strains (formae specialis) of *Erysiphe graminis* which occurs world wide. Yield losses can be as much as 20% in both wheat and barley. Appressoria form on the leaf surface and penetration results in the formation of haustoria in the epidermal cells. There may be visible growth of the white mycelium and conidia on the leaves after only a few days and spread can be rapid because of the large number of spores produced. As with the rusts the photosynthesis is reduced with similar effects and control is again by resistant varieties and fungicides such as tridemorph. There are also some quite specific fungicides for mildew such as ethirimol which may be used as a seed dressing or a spray, and many more general fungicides such as triadimefon.

Finally, with cereals, we might mention a necrotroph, leaf blotch of barley caused by *Rhynchosporium secalis* which is common in cool maritime climates. Yield losses from this trash-borne disease (p. 139) may be of the order of 20 to 30%. There are some varieties with a little resistance, but the fungicides used for mildew and rust also give control and are usually used as a combined spray of captafol plus ethirimol or tridemorph plus carbendazim.

Leaf spots of other plants include the familiar *Diplocarpon rosae* which is all too common in many gardens, and has recently become more common since clean air legislation has reduced sulphur dioxide levels near urban centres. *Venturia inaequalis* causes apple scab and has already been mentioned because of its major importance on fruit (p. 39) but it also forms a raised brown or black lesion on the leaves and overwinters on dead leaves on the ground. Perithecia are produced in spring and the ascospores infect the flowers, developing fruit or the leaves. The ascospores need water to germinate and penetrate in a time which depends on the ambient temperature; 28 h are needed at 6°C but only 12 h at 26°C so the disease is favoured by warm, wet springs. Protectant sprays are used, especially after rainy periods.

An entirely different sort of disease, peach leaf curl, is caused by *Taphrina deformans*, an obligate biotroph that does not form haustoria. The fungus overwinters on the bark and in buds (p. 54) and infects the developing leaves where it grows beneath the cuticle and intercellularly in the mesophyll (Fig. 4.9 B). The mesophyll undergoes hypertrophy and hyperplasia to produce a thickened, distorted leaf with a reddish or purple colour (Fig. 4.9 A and C). Asci form under the cuticle and eventually burst through it to release the ascospores. Control depends on early application of protectant fungicides, by the time the leaf is visibly distorted it is too late to control the disease that year.

Bacterial leaf diseases are not usually very important, though they may be locally serious on particular crops they do not have the world-wide impact of the fungal cereal diseases just discussed. Wild-fire of tobacco is caused by *Pseudomonas tabaci* which produces a host specific toxin, tabatoxin. In seedling tobacco plants the leaves have an advancing wet rot with a water soaked margin. In the field yellowish spots on the leaf develop into brown necrotic lesions with a chlorotic halo and the centre eventually dries out and falls away. Lesions may merge to cover almost all the leaf, making the crop useless and killing the plant in severe cases. There are resistant cultivars and copper and streptomycin sprays are sometimes used in nurseries. There is no effective chemical control on a field scale. Leaf blight of rice is another

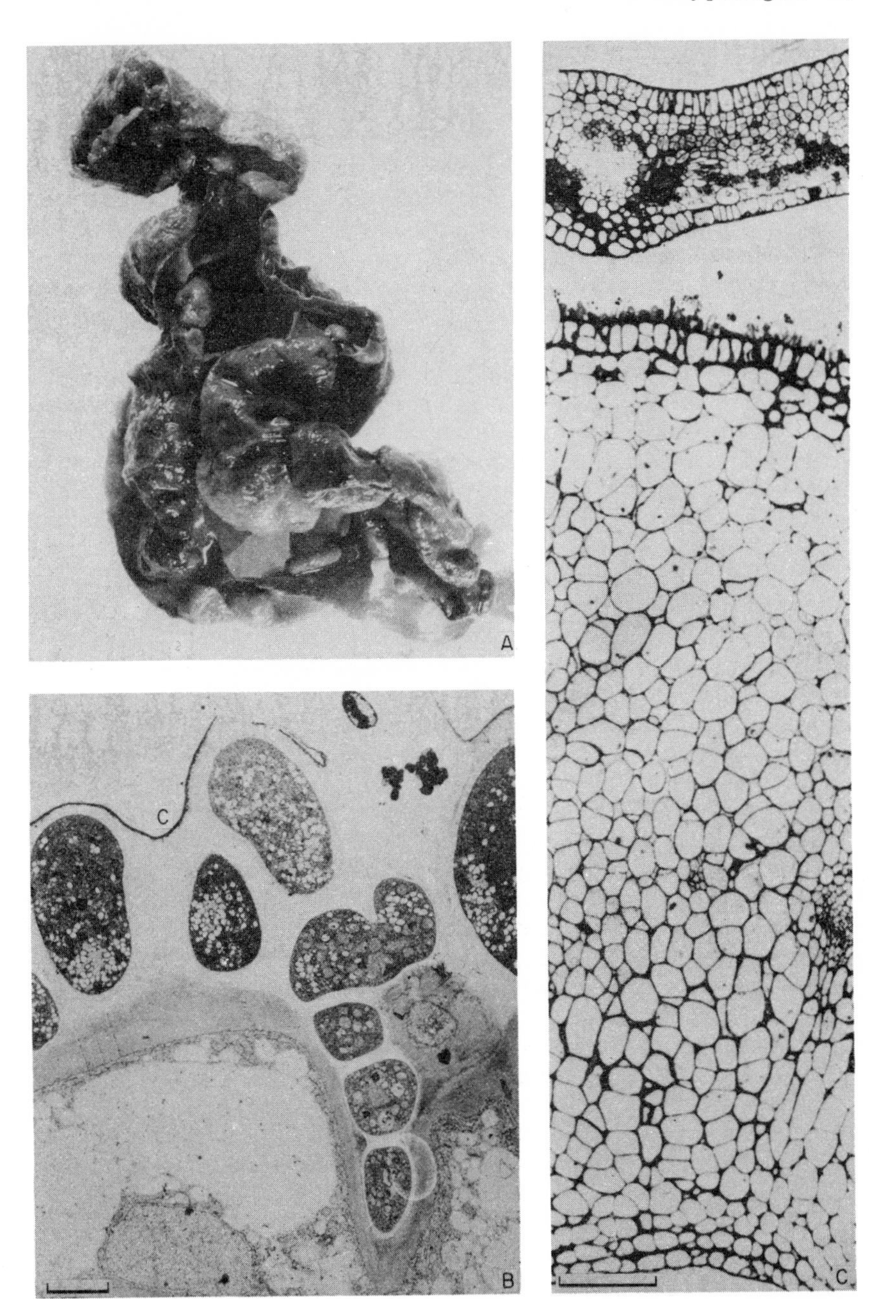

Fig. 4.9 *Taphrina deformans*. **A.** General view of a distorted leaf. ½ x natural size. **B.** Transmission electron micrograph to show young asci and ascogenous hyphae under the cuticle (C) with penetration between the epidermal cells. Bar = 5 μm. **C.** Light micrographs, transverse sections of the leaf at early infection (top) and late infection (below), with cell division and enlargement (hyperplasia and hypertrophy) giving a greatly thickened leaf. Each section at the same magnification. Bar = 100 μm. (Photographs courtesy of M. Syrop and A. Beckett, Department of Botany, University of Bristol.)

important bacterial leaf disease (caused by *Xanthomonas oryzae*), which may result in losses of up to 60% in India and Indonesia.

Finally, in this consideration of pathogens on leaves we must discuss viruses, which produce many diseases of agricultural importance. The classification of viruses and the general symptoms that they produce have already been mentioned (p. 4). Viruses in leaves may be systemic, having been carried over in the seed or been introduced by grafting or in vegetative propagation from infected stock plants. There are cases of plant virus diseases with no known vector, but with a highly susceptible plant which is easily wounded and carries sap that is heavily infected. Potato X virus occurs in the sap released from damaged leaf hairs as leaves rub against one another in the wind. Virus is transferred from wound to wound. The main vectors of leaf infecting viruses are however aphids, leaf hoppers and mites. These are all sap suckers which penetrate the leaves with stylets. They may suck up virus particles and inject them again when they next visit a plant of that species. The virus may be mechanically carried in the mouth parts but does not usually remain viable for more than a few hours. The virus may also be taken into the gut and from there pass into the haemolymph and thence to the salivary glands, to be re-injected after multiplication. Finally, the virus may live, and probably multiply, in its vector for many days or even weeks: it has a plant and an animal host in which it can multiply. Times of persistence and the latent periods vary greatly between viruses and vectors: some examples are listed in Table 4.3.

Once inside the host protoplast the virus particle is 'uncoated'; the protective proteins are removed to expose the nucleic acid (usually RNA, Table 1.2). This is replicated by the host enzymes and virus proteins are also produced by the host, both using the information in the viral nucleic acid. The virus particles are then assembled and this may be an automatic process.

Quite how the virus produces the symptoms in the plant is not clear. Obviously at the ultrastructural and biochemical level the metabolism is dictated by the virus synthesis and there are often changes in the chloroplast structure and products, increases in the respiration rate, and changes in protein and nucleic acid synthesis. There is however no general account of why one virus produces a mottling and another a stripe on leaves, or even how different symptoms in different plants may be caused by the same virus. Virus diseases are not like the bacterial and fungal diseases just discussed, which are usually quite specific. Though a particular virus may have a name such as tobacco ring spot it does not cause tobacco ring spot only: this is an extreme example, because this particular virus is known to infect at least 246 species of higher plant in 54 different families and it produces different symptoms on different hosts!

There may be interactions between viruses in leaves and other pathogenic micro-organisms. For example leaf roll viruses may give higher humidity or retain free water for longer on the leaf surface and this has been shown to increase late blight of potatoes (*Phytophthora infestans*) and chocolate spot of beans (*Botrytis fabae*) both of which need free water for infection. There are also other associations which do not depend so obviously on microclimatic effects; mild yellows virus of beet increases the frequency of *Alternaria* infections. Though there is no direct evidence it seems probable that leaf viruses could affect the saprotrophs in similar ways

Table 4.3 Characteristics of the transmission of some leaf viruses. (Selected from Gibbs, A. & Harrison, B., 1979, *Plant Virology; the principles*. Edward Arnold, London. pp. 292.)

Vector	Virus	Minimum time to acquire virus	Minimum latent period before re-injection	Minimum time to inoculate virus	Maximum persistence in vector	Multi-plication in vector
Myzus persicae (Aphid)	Potato Y	10 s	0	15 s	2 h	No
Bemisia tabaci (White fly)	Tomato yellow leaf curl	30 min	21 h	30 min	20 days	Not known
	Cucumber vein yellowing	20 min	0	10 min	6 h	No
Hyperomyzus lactucae (Aphid)	Sow thistle yellow vein	2 h	8 days	< 1 h	weeks	Yes
Circulifer tenellus (Leaf hopper)	Beet curly top	1 min	4 h	1 min	weeks	Unlikely
Agallia constricta (Leaf hopper)	Wound tumor	< 1 h	12 days	< 1 h	weeks	Yes

Control of plant diseases on leaves

Detailed consideration of this enormous topic is outside the scope of the present text and it is well treated in many plant pathology texts (Dickinson & Lucas, 1982; Manners, 1982) and specialist books (Jenkyn & Plumb, 1981; Staples & Toenniessen, 1981). We have mentioned some specific examples in the discussion of particular diseases, but it will be considered again here particularly from the point of view of control which involves microbial interactions on the leaf.

Disease is the exception; most plants do not get most diseases. There is immunity in plants (p. 26) and variation in the resistance of different cultivars (p. 23). The exploitation of this host resistance in plant breeding programmes has been particularly used for leaf diseases, especially of cereals. However single-gene resistance may soon break down and it is a race for the plant breeder to keep ahead of the changing pathogens. If a new cereal variety retains its resistance to several pathogens for more than a few years then it is doing well. Nevertheless plant breeding remains a major line of defence against leaf diseases.

Another major defence is good agricultural practice. This means using rotation of crops to reduce the inoculum level from carry over of dead leaves on the ground and general good hygiene over crop residues. The latter is particularly important with some leaf diseases. Hygiene might also include the removal of alternate hosts of some rusts so as to break the disease cycle.

Failing all else pesticides are used to control leaf diseases. There are an enormous number of active chemicals and many more different formulations in use. The Ministry of Agriculture Fisheries and Food in Britain list more than 75 formulations of fungicides approved for use on cereals alone (MAFF, Use of Fungicides and Insecticides on Cereals, 1983. ADAS Booklet 2257(83)). These fungicides may be protectant in that they provide a toxic deposit on the leaves and kill germinating spores as they start to grow or they may be systemic, being translocated, through the xylem usually, and providing some control within the leaf after infection.

Another possible method of controlling leaf diseases is by biological means, using some other micro-organism(s) to kill or discourage pathogens (Blakeman & Fokkema, 1982; Blakeman, 1982). There are many examples of this on leaves known from experimental work in the laboratory, but they are not used commercially. This is largely because of the range of relatively cheap and effective fungicides available. Even if the phenomenon of biological control is not actively used on leaves, there is quite a lot of evidence that it occurs naturally and reduces or eliminates disease, without being noticed since serious disease does not appear. Thus diseases caused by *Cochliobolus sativus* on leaves may be made more serious by the application of benomyl-containing fungicides: these are known not to affect *C. sativus* but they do kill saprotrophs and other pathogens. Removal of competition leads to greater growth of *C. sativus* and worse disease than before the spray, and the implication is that *C. sativus* is normally partly controlled by the microflora.

Nutrient competition occurs on the leaf surface between bacteria and fungal spores and between fungal spores and yeasts. A reduction in the surface growth of pathogens with reduced nutrients has also been shown. Even simple leaf washing for one hour, which removes most leaf surface sugars, reduced production of hyphae by *Erysiphe cruciferarum* (Table 4.4). There are however more subtle interactions; the nutrients may be removed by antagonistic saprotrophic phylloplane flora. The spore of *Botrytis*, for example, may leak nutrients as it hydrates prior to germination and these nutrients are subsequently taken up again for growth. Antagonistic bacteria or continuous leaching may however remove the nutrients and reduces infection. Bacteria may also reduce general nutrient levels on the leaf.

Other forms of antagonism are also possible on the leaf surface. Antibiotics may be produced and hyphae or spores lysed. It is possible that non-pathogens,

Table 4.4 Germination of *Erysiphe cruciferarum* in relation to leaf washing. (From Purnell, T.J., 1971, in Preece, T.F. & Dickinson, C.H.)

	Total carbohydrate (μg cm^{-2})	% of spores with primary hyphae				% spores with primary and secondary hyphae			
		Expt. 1	2	3	4	1	2	3	4
Unwashed leaf	0.321	70	36	42	43	30	10	16	0
Washed leaf	0.050	30	15	9	23	0	0	3	0

or avirulent races of pathogens, may stimulate the host defences and so 'prepare' the plant to resist the eventual serious attack.

Biological control may also involve mycoparasitism: the parasitism of one fungus by another. The host fungus may be lysed, physically excluded from developing fruiting bodies or have the parasitic hyphae coiled around it so restricting growth and development (Deacon, 1983). There are more than 80 fungi known to parasitize rusts and powdery mildews (Kranz, 1981) and one fungus, *Darluca filum*, is known to parasitize over 360 rust species. Many of the mycoparasites show this lack of specificity and, like *D. filum*, occur mostly in the tropics and sub-tropics where there may be considerable potential for such biological control because of the lack of technology and money to make or buy fungicides.

Effects of pesticides on leaf surface microbes

Knowledge of these effects is unfortunately confined to temperate agricultural crops under intensive western farming systems, even though pesticides are more widely used. The response of the micro-organisms depends very much on the particular chemical used and on the resident microflora. Obviously fungicides tend to kill fungi, some are very broad spectrum but others are less so; benomyl does not affect zygomycetes, Mastigomycotina and some particular genera such as *Alternaria*. A few fungicides are very specific: tridemorph affects cereal mildews and not other 'non-target' fungi. It is therefore possible to change the balance of fungi by selectively inhibiting particular genera and so allowing competitors to grow. The importance of different diseases may change as one is reduced by a spray, only to allow expression of another which is not affected by that fungicide and which had previously not been considered important. This is only considering the effects of fungicides on fungi; what effects do bacteriocides and insecticides have on fungi? Spray programmes can be very complex; in the data shown in Fig. 4.4 the apple trees had 14 applications of fungicide, 5 of insecticide and 2 of bacteriocide in the one growing season. There is usually a decrease in the fungal, yeast and bacterial populations of several orders of magnitude in the first year, and the effects may persist at a less drastic level, into the second year. Species diversity is also reduced by fungicide sprays. Insecticides may reduce fungal disease, often by unknown mechanisms, though the elimination of vectors is an obvious possibility. There is little direct evidence for deleterious effects of commercial herbicides on the leaf surface flora, most attention having been paid to the microbiology of the dead leaves on the soil rather than to the effects on non-target plants of selective herbicides. Herbicides extensively used on cereals, for example, kill the broad-leaved weeds, but what happens to the cereal microflora? Some studies have been done in culture with media ammended with herbicide and they suggest that there may be some effect on individual species. Furthermore some selective herbicides may cause slight phytotoxicity to other plants and allow development of unspecialized opportunistic necrotrophs. Herbicides applied to leaves may be translocated to roots and affect soil organisms. Effects on the complex communities and their interactions on the leaf surface itself are not known.

Effects of pollutants on leaf surfaces

The main leaf surface pollutants, apart from pesticides just discussed, are acid rain, gaseous pollutants like sulphur dioxide, hydrogen sulphide, nitrous oxides, carbon monoxide and fluorine, and metals such as zinc and lead which are usually particulate. Much work has now been done on car exhaust gases, especially of cars using leaded petrol. The exhaust gases, under the action of ultra-violet light, may form ozone, peroxyacetyl nitrate (PAN), aldehydes, oxides of nitrogen and ethylene.

Most of the information is restricted to plants growing in the developed countries where the pollutants are produced. No doubt such pollutants will eventually become important in the tropics but since we know almost nothing about the normal flora here (p. 63) it is not possible to even speculate about what effects pollutants might have. There are suggestions that higher temperatures and humidity increase breakdown rates.

There are several conflicting ways of looking at this problem, for some of the pollutants just listed are considered helpful in the right place at the right time. Thus some copper and zinc compounds are used as protectant fungicides on leaves, sulphur dioxide is used to kill wild yeasts in fermentations and as a preservative in some foods and ozone is used to control spoilage of some fruits and vegetables during storage. These chemicals work by reducing microbial populations and this is what they, and many other chemicals, do on leaves.

The mechanism of pollutant effects is very varied. Some dusts may cause abrasion damage to leaves or block stomata. The resultant leaf necrosis can encourage saprotrophic microbes, if they are not themselves adversely affected, and may cause increases in some diseases (e.g. leaf spot of beet caused by *Cercospora beticola*). Gases may be actively toxic as they are or they may interact with other factors in the environment. Sulphur dioxide is toxic but it also forms sulphuric acid with rain. Acid rain may have a direct pH effect to discourage microbes or it may cause other toxic compounds like heavy metals to come into solution and therefore be more readily available than as insoluble dust particles. There are thus synergistic and competititive effects of pollutants on leaves: the combined effects of acid rain and heavy metals may be greater than the sum of their individual effects, but in the presence of zinc the toxicity of cadmium is reduced because the less toxic zinc competes successfully for the transport sites on membranes, reducing the uptake of the more toxic cadmium. Pollutants, from car exhausts or industrial complexes for example, rarely occur singly and it is these interactions that make the understanding of pollutant effects on micro-organisms so difficult.

There may also be interactions with other, non-microbial components of the ecosystem. It has been shown that heavy metals can lead to the build-up of leaf litter in woodlands (Martin & Coughtrey, 1982). The first possibility is that the leaves are not decayed by microbes because of their metal content or that the resident microbial population before leaf fall was altered. However the micro-organisms present seem to be similar in numbers and species to normal leaves and litter, though many of them are heavy-metal tolerant strains. The main difference is that heavy metals reduced or removed the earthworms and terrestrial isopods (crustaceans, e.g. woodlice) from the litter and the activity of these animals is very important in increasing microbial turnover, and hence decay rates, in litter.

Having made these generalizations there are two pollutants whose effects on microbes on leaves have been studied in enough detail to be worth considering separately. These are sulphur dioxide and heavy metals (Heagle, 1973, for effects on pathogens particularly).

Sulphur dioxide has been shown to decrease the incidence of many leaf pathogens including *Puccinia graminis*, various rusts on conifer needles, needle cast of larch and apple scab (p. 68). The response of tar spot on sycamore (caused by *Rhytisma acerinum*) is predictable enough to be used as a pollution indicator. It was very rare in urban areas but has recently become more common (see also p. 68 for *Diplocarpon rosae*). There is no tar spot at mean

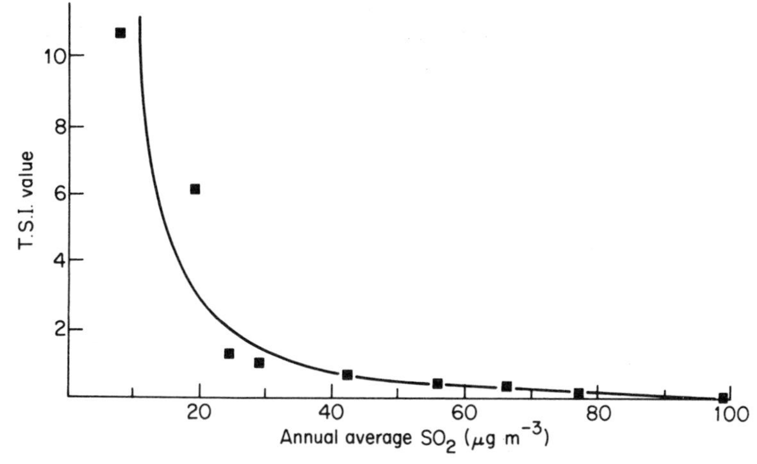

Fig. 4.10 Relationship between the annual average sulphur dioxide concentratons and tar spot index (TSI). (From Bevan, R.J. & Greenhalgh, G.N., 1976, *Environmental Pollution*, **10**, 271–85.)

sulphur dioxide levels above 85 $\mu g\ m^{-1}$ and it increases as the sulphur dioxide levels drop (Fig. 4.10, see also Table 5.1 for lichens).

However some saprotrophic fungi or weak necrotrophs, which are known to be sulphur dioxide tolerant, may be favoured as their competitors are killed or reduced. Needle cast of spruce has been shown to increase on exposure of the host to sulphur dioxide, but the reason is unknown. There are occasional references, in studies on diseases, to the reduction in the general saprotrophic flora of leaves (e.g. in disease caused by *Hysterium*) but no detailed work. This is another hint of the saprophyte/pathogen interactions involved in natural biological control which only show when the balance is disturbed (p. 72). Yeasts on leaves are presumably reduced by sulphur dioxide in view of their reaction in fermentations, but there are no data specifically for leaves. There are no data for the effects of sulphur dioxide on bacteria on leaves, though it is toxic in culture.

Culture work with sulphur dioxide shows that spores of fungi are more sensitive than the mycelium. Sclerotia are the most resistant structures to sulphur dioxide. There is also an interaction with water, for high humidity and surface wetness increases sensitivity to the gas. This is probably equivalent to acid rain effects which can produce separate effects to sulphur dioxide *per se*. Acid rain causes a reduction in rust of oak leaves (*Cronatium fusiforme*), rust of beans (*Uromyces phaseoli*), halo blight of beans (*Pseudomonas phaseolicola*), and diseases of vines.

Effects of heavy metals on leaf surface fungi are of two sorts, those associated with general urban pollution of leaves and the very high levels near to special sources of metals such as smelters.

Different fungi and bacteria show different degrees of tolerance to available heavy metals and even within species there are tolerant and sensitive strains. *Aureobasidium*, *Cladosporium* and *Epicoccum*, perhaps the commonest leaf surface microbes, often have highly tolerant strains and may indeed increase in abundance under heavy metal pollution (*Aureobasidium* with lead, Table 4.5). Yeasts in general do not seem to be very tolerant, with the exception of *Cryptococcus*. Bacteria, especially orange and yellow pigmented ones which are common on leaves, can tolerate quite high environmental levels of metals and may be positively correlated (Table 4.5).

Table 4.5 is only part of a wider study of the effects of heavy metals on micro-organisms, but it shows several interesting points. Zinc levels decrease microbial populations though most of the effect is due to the inevitable combination of the dust with other metals, zinc alone only causing a reduction in non-pigmented yeasts and yellow bacteria. Populations of *Aureobasidium*

Table 4.5 Total and partial correlation coefficients between the levels of heavy metals and the numbers of micro-organisms on leaves of Hawthorn. Negative correlations indicate that organisms are decreased by the presence of metal: positive correlations indicate that an organism increases in response to the metal, i.e. is tolerant and exploits the lack of competition. The partial correlations have the effects of the other two metals removed. (From Bewley, R.J.F. & Campbell, R., 1980, *Microbial Ecology*, **6**, 227–40.)

Micro-organisms	**Zinc** Total	Partial	**Lead** Total	Partial	**Cadmium** Total	Partial
Aureobasidium pullulans	− 0.62**	− 0.24	− 0.54*	+ 0.62**	− 0.64**	− 0.53*
Cladosporium	− 0.52*	− 0.13	− 0.48*	+ 0.18	− 0.52*	− 0.15
Total filamentous fungi	− 0.62**	− 0.18	− 0.57**	+ 0.23	− 0.61**	− 0.18
Sporobolomyces roseus	− 0.31**	− 0.14	− 0.27**	+ 0.14	− 0.34***	− 0.14
Non-pigmented yeasts	− 0.93***	− 0.54*	− 0.88***	+ 0.53*	− 0.92***	−0.45
Pigmented yeasts	− 0.53*	− 0.15	− 0.50*	+ 0.11	− 0.52*	− 0.08
Orange pigmented bacteria	− 0.71***	+ 0.25	− 0.78***	− 0.72***	− 0.69***	+ 0.67***
Yellow pigmented bacteria	− 0.87***	− 0.68***	− 0.82***	− 0.13	− 0.80***	+ 0.49*
Total bacteria	− 0.79***	+ 0.38	− 0.87***	− 0.66**	− 0.82***	+ 0.13

Significant correlations are * at the 5% probability level; ** at 1%; and *** at 0.1%.

are generally reduced by all metals except lead to which its numbers are positively correlated. The same is true of non-pigmented yeasts. *Sporobolomyces*, a pink yeast, is affected adversely by metals and is sensitive to sulphur dioxide; it does not occur within 2 km of the smelter, but mostly because of pollutants other than metals (see the low correlation coefficients with metals). Filamentous fungi and 'total' bacteria are reduced by all three metals. The bacteria are especially sensitive to lead but the orange and yellow chromogens are tolerant to cadmium. This means that total populations are usually decreased by metals but particular species, genera or higher taxa develop tolerance and may increase in numbers. The balance of different organisms within the community is thus changed. It is therefore very difficult to predict what effects metal will have on a community, and there may be complications such as the effect on the soil animals noted above (p. 74).

In general the effects of pollutants on the leaf surface flora are rather poorly understood. We have a few sets of data for particular pollutants, but even these are scattered geographically and are for different climatic conditions. We have no overall view of what atmospheric pollutants are doing, or may do, to either saprotrophs or pathogens on leaves.

Conclusions

Our knowledge of leaf microflora is very uneven. We know a great deal about pathogens of most temperate agricultural crops, because of their economic importance to western countries. We have a reasonable knowledge of the saprotrophic floras of such plants, which has often been acquired as a side effect of the pathogen studies. The microflora of natural vegetation is almost unknown. Forests are represented by a few studies, mostly in relation to disease or by the intensive work on a few trees. For the whole of the tropics, both agricultural and natural vegetation, there is virtually no information except that some particular fungus has been described as living on this or that leaf, or that some pathogen occurs on this host species. The pioneer work of Ruinen (1961), excellent though it was, remains almost the sole source from which generalizations are made about 'the tropics', though she did most of her work in a small part of south-east Asia. The only consolation about all this is that when investigations are made, the leaf flora seems to be remarkably similar within a climatic or vegetation type. The biomass, except in the tropics, seems quite low so perhaps leaf saproptrophs do not warrant more attention than they at present get in research budgets. We do not however know much about the activity of the biomass and it remains an area which needs more study. Presumably most of the work will continue to be carried out on agricultural crops, and this is justified on the grounds of economics and food production. If pesticides are to continue to be used on crops we need to know more about their effects on leaf surface microbes. Alternatively if the use of pesticides is to be reduced for ecological reasons (though there is no prospect of this in western agriculture) then we need to turn to biological and integrated control measures for which a great deal of understanding is needed of microbial interactions on the leaf surface. In any event there is a continuing need to study leaf microbiology.

Selected references and further reading

Andrews, J.H., Kenerley, C.M. & Nordheim, E.V. (1980). Positional variation in phylloplane microbial populations within an apple tree canopy. *Microbial Ecology* **6**, 71–84.

Blakeman, J.P. (Ed.) (1981). *Microbial ecology of the phylloplane*. Academic Press, London. pp. 502.

Blakeman, J.P. (1982). Phylloplane interactions. In Mount, M.S. & Lacy, G.H. (Eds) *Phytopathogenic prokaryotes*. Academic Press, New York. p. 307–33.

Blakeman, J.P. & Fokkema, N.J. (1982). Potential for biological control of plant diseases on the phylloplane. *Annual Review of Phytopathology*, **20**, 167–92.

Deacon, J.W. (1983). *Microbial control of plant pests and diseases*. van Norstrand Reinhold, Wokingham, England. pp. 88.

Dickinson, C.H. (1982). The phylloplane and other aerial plant surfaces. In Burns, R.G. & Slater, J.H. (Eds) *Experimental microbial ecology*. Blackwell Scientific Publications, Oxford. p. 412–30.

Dickinson, C.H. & Lucas, J.A. (1982). *Plant pathology and plant pathogens*, 2nd edition. Blackwell Scientific Publications, Oxford. pp. 229.

Dickinson, C.H. & Preece, T.F. (Eds) (1976). *Microbiology of aerial plant surfaces*. Academic Press, London. pp. 669.

Heagle, A.S. (1973). Interactions between air pollutants and plant parasites. *Annual Review of Phytopathology*, **11**, 365–88.

Heitfuss, R. & Williams, P.H. (Eds) (1976). *Physiological plant pathology*. Encyclopedia of Plant Physiology, N.S. Vol. 4. Springer Verlag, Berlin. pp. 690.

Jenkyn, J.F. & Plumb, R.T. (Eds) (1981). *Strategies for the control of cereal diseases*. Blackwell Scientific Publications, Oxford. pp. 219.

Juniper, B.E. & Jeffree, C.E. (1983). *Plant surfaces*. Edward Arnold, London. pp. 93.

Kranz, J. (1981). Hyperparasitism of biotrophic fungi. In Blakeman J.P. (Ed.) *Microbial ecology of the phylloplane*. Academic Press, London. p. 327–52.

Manners, J.G. (1982). *Principles of plant pathology*. Cambridge University Press, Cambridge. pp. 264.

Martin, M.H. & Coughtrey, P.J. (1982). *Biological monitoring of heavy metal pollution: land and air*. Applied Science Publishers, London. pp. 475.

Preece, T.F. & Dickinson, C.H. (Eds) (1971). *Ecology of leaf surface micro-organisms*. Academic Press, London. pp. 640.

Rhodes-Roberts, M.E. & Skinner, F.A. (Eds) (1982). *Bacteria and plants*. Academic Press, London. pp. 264.

Round, R.E. (1981). *The ecology of algae*. Cambridge University Press, Cambridge. pp. 653.

Ruinen, J. (1961). The phyllosphere: an ecologically neglected milieu. *Plant and Soil*, **15**, 81–108.

Scott, P.R. & Bainbridge, A. (Eds) (1978). *Plant disease epidemiology*. Blackwell Scientific Publications, Oxford. pp. 329.

Staples, R.C. & Toenniessen, G.H. (Eds) (1981). *Plant disease control*. Wiley, New York. pp. 339.

5
Microbiology of Stems

Introduction

There is an enormous variety of habitats included here, from temperate grass and herb stems and coniferous and broad-leaved trees to tropical vegetation of all types. The surface texture, chemical composition and availability of water varies greatly with species and climate. However these complications will not be serious because there is so little information on the microbiology of stems, apart from wood decay, that comparisons of different climates or species will not usually be possible. The corollary of this is that it should be remembered that many of the generalizations made will be based on few data.

The initial discussion of herbaceous stems and bark will centre on epiphytes which do not degrade the substrate to any extent. Within woody stems the overwhelming amount of information is on wood decay and we will consider this briefly. As in previous chapters, the decay of litter on the soil will not be considered extensively; as soon as the material ceases to be readily identifiable as a stem then it passes out of the scope of this chapter (but see p. 153 – 170).

The methods used in these studies are really quite simple. Direct examination may be at several levels. Electron microscopy has occasionally been used but the light microscope hardly at all because most of the material is opaque. Various washing, scraping or swabbing techniques produce suspensions of organisms for examination in the light microscope. There are some studies of sectioned material, building up biomass or biovolume figures from measured areas on a sample or on serial sections. Direct examination, however, usually involves looking at the herbaceous stem, possibly after incubation in a damp chamber, with a dissecting microscope or a hand lens to identify small fungal structures, or even just looking for macroscopic food bodies, especially of Aphyllophorales on large stems. These approaches, using fruiting bodies, give no information on the actual distribution of the fungal mycelium, which is the main part of the life cycle.

Culture methods are used for surfaces by washing and plating, and for larger, solid structures by picking out small pieces of wood and plating these onto media. Basidiomycetes which are important in wood decay usually grow slowly in culture and may not compete on dilution plates with heavily sporing, fast growing deuteromycetes.

Combinations of methods involving careful dissection of stems, stumps or even whole trees has been done by cutting the wood into thin slices, plotting the distribution of rot types and culturing to identify the fungi and to do comparative laboratory tests on the different strains or mating types. The three-dimensional distribution of organisms can then be reconstructed and this yields a lot of information on the interactions between organisms and on successions in decaying stems.

The information available from the use of these methods is heavily biased towards plant pathology and therefore towards fungi. Hardly anyone has gone out just to see what is there: almost always the studies are concerned with what has killed this tree, or what is rotting this timber. Saprotrophs are usually noted incidentally, if at all, and most bacteria are not identified; they may occasionally be counted but usually are discarded as 'contaminants'. Protozoa have not even been thought about except for a few anecdotal comments and unpublished studies, though they seem to be abundant. Algae are of course confined to the stem surfaces though only rarely does anyone bother to find out what they are. Lichens come in for quite a lot of investigation, especially in relation to pollution to which they are so especially sensitive that they have been used as indicators of sulphur dioxide levels in the air.

The surface of herbaceous and other small stems

Green herbaceous stems seem to have a very similar flora to leaves, though there is almost no detailed information. As such stems senesce and die they are colonized by a range of fungi, usually starting at the stem tip and working back as the plant dies down for the winter, in temperate countries. *Urtica* sp. (Urticaceae, stinging nettle), *Heracleum* sp. (Umbelliferae, hog-weed), *Pteridium* (bracken, a fern) and also some grasses (*Agropyron, Dactylis*) are colonized as soon as they senesce by deuteromycetes such as *Cladosporium, Aureobasidium pullulans, Alternaria* and *Epicoccum purpurescens*. As decay proceeds and the stems reach the ground there is colonization by ascomycetes such as *Pleospora* and *Leptosphaeria* and later some basidiomycetes such as *Corticium*.

There are many pathogens of agriculturally-important crops which attack the stem, perhaps the best known of which is black stem rust of wheat (*Puccinia graminis*). This is a most important disease, not so much because of the direct losses it causes but more because of the money and effort involved in breeding resistant varieties and in fungicide programmes to control it.

Another major stem disease, caused this time by a bacterium, *Agrobacterium tumefaciens*, is crown gall of a very wide variety of plants (1193 species), including many with woody stems, though these are usually attacked when young in nursery beds so we will consider them here with the herbaceous ones. A large growth or tumour forms on the stem, usually at ground level, though it may also occur on roots and branches. The most seriously affected crops are peaches and plums, apples, pears, roses and sugar-beet; losses run into many millions of pounds per annum worldwide. Infection is normally through wounds and the conditions of infection and disease development are well-known (Giles & Atherley, 1981). Control depends on hygiene, crop rotation and the avoidance of wounding. Treatment with antibiotics such as terramycin and aureomycin is effective though expensive. There are also biological methods using avirulent strains of the bacteria.

Apart from its agricultural importance there is interest in using the disease to study other matters. It has been considered in cancer studies in the hope that this 'plant cancer' may throw some light on human diseases. However the most interesting aspect in recent years has been in the study of molecular genetics and genetic engineering. The bacterium is pathogenic because it contains the *Ti* plasmid which is transferred to the host and incorporated in its genome. This is remarkable in that it is transfer of genetic information from a prokaryote to a eukaryote. This is now being investigated in great detail (Roberts, 1982), for though it is relatively easy to tranfer DNA amongst bacteria, and there are very well developed methods for genetic engineering in prokaryotes, this is not true of eukaryotes. *Agrobacterium tumefaciens* is therefore a possible link to transfer prokaryotic DNA, possibly itself artificially introduced into a bacterium in previous manipulations, into a eukaryotic plant by using the *Ti* plasmid. There are many problems yet to be overcome and it is not even certain that the long-term aim of being able to introduce prokaryotic DNA into eukaryotes is going to be useful, but the system is well worth investigating. For example it may be possible to introduce nitrogen-fixing genes (*nif*) into higher plants or to enable nitrogen-fixing symbionts such as *Rhizobium* to form effective nodules, on say wheat, by introducing into the new crop the appropriate genes from legumes for the lectin recognition system (p. 126). The new symbiosis could help reduce nitrogen fertilizer use, but what the effect would be on wheat yield is uncertain because photosynthate is diverted to supply the rhizobia with nutrients (p. 147).

There are few virus diseases of stems, at least few that have been recognized, except the general growth distortion or the stunts which operate through the alteration of hormone levels.

There are of course many atypical or specialized herbaceous stems forming storage organs or underground rhizomes and similar structures. The surface saprotrophic flora of these structures has been little investigated but it seems to be similar to roots (p. 110), gaining micro-organisms from the soil and being adapted to growth in this environment. There are some very important underground stems which are agricultural crops, especially the potato. The post-harvest losses of this crop have been noted already (Table 3.1). Such rots include fungi and bacteria (e.g. *Erwinia*) and the potato is prone to a large number of virus diseases, many of which can be carried in the tuber as well as being arthropod-transmitted on the above-ground parts. The outstanding disease of potatoes is late blight caused by *Phytophthora infestans*. This fungus infects the leaves and can kill the plant but it can also rot the stem tuber. The infection may be latent, only appearing during storage and causing serious rotting. Apart from the loss of the crop, the infected tubers also provide a source of inoculum for further infection. The Irish potato famine, caused by this fungus, has been written about on numerous occasions as an example of a devastating fungal disease, being responsible for millions of deaths and the enforced migration of millions more people to form a considerable proportion of the American population. This is an interesting disease from many aspects. For example much work has been done on its epidemiology. It spreads in two ways, firstly by air-borne sporangia and secondly by free-swimming zoospores. The former give long-range spread with odd outbreaks in the middle of the field and then there is a cluster of infection sites around this initial infection focus, especially in wet, humid weather which allows free water for the zoospores to swim. This sort of epidemiological information is used to predict high risk periods in terms of the amount of rain, the relative humidity and the temperature which will allow zoospore release and movement. If suitable conditions prevail a protectant spray can be applied, to the leaves, in advance of any symptoms being visible so preventing a potential epiphytotic at its earliest stages.

Another potato disease which is very common is scab, caused by *Streptomyces scabies*, one of the few agriculturally-important actinomycetes. Chemical means of control are often ineffective below soil level and anyway cheap bacteriocides are much less available than fungicides. Control can be obtained by ploughing in a green manure crop, such as alfalfa, or on a gardening scale by putting organic matter in the planting trench. This is thought to encourage general microbial activity in the soil which is antagonistic to the pathogen, reducing disease by biological means rather than by applying chemicals or breeding resistant varieties.

The surface of woody stems – bark

The conditions on herbaceous stems are very different as soon as secondary thickening and the bark formation occurs. In mature tree stems bark is about 12% of the volume, though it varies with species. The surface of the stem becomes waterproof and contains toxic and water-repellent substances. There are also many bark types, some macroscopically smooth such as that of beech (*Fagus*), but they may be ridged, containing crevices and cracks which are protected to some extent from the desiccation and higher temperatures of the true bark surface. Bark is continuously replaced from below to allow for increased stem growth, and the bark surface is worn away or flakes off: organisms are liable to be removed from this habitat unless they have some means of redistributing themselves or of insuring a continuous supply of inoculum. Bark is thus an inhospitable growth substrate in many ways, it is often dry and subjected to insolation, it is not high in available nutrients and it contains toxic tannins, terpenes and polyphenols. There are not many microbes living actually in the bark, it is mostly a surface community based on primary production by algae and lichens or in specialized, usually pathogenic conditions on a nutrient rich slime flux from within the tree.

Actually in the bark there are only a few saprotrophic imperfect fungi, especially sooty moulds in the Capnodiaceae and common genera such as *Penicillium* and *Trichoderma*. Some saprotrophic ascomycetes are also recorded and there are pathogens such as *Nectria cinnabarina* which causes coral spot (Fig. 5.1), usually on suppressed branches, though it can girdle the stem as it forms a canker, so killing the branch above the girdle. There are other *Nectria* diseases of bark, especially beech bark disease caused by *N. coccinea* which is invariably associated with a scale insect, *Cryptococcus fagi*. The latter provides wound sites and generally weakens the tree. Again in more severe attacks the tree may be girdled or individual branches may die back.

A similar effect is produced on sycamore suffering from sooty bark disease which is caused by the deuteromycete *Cryptostroma corticale* which grows in and under the bark, killing it so that it falls off to reveal the sporulating fungus.

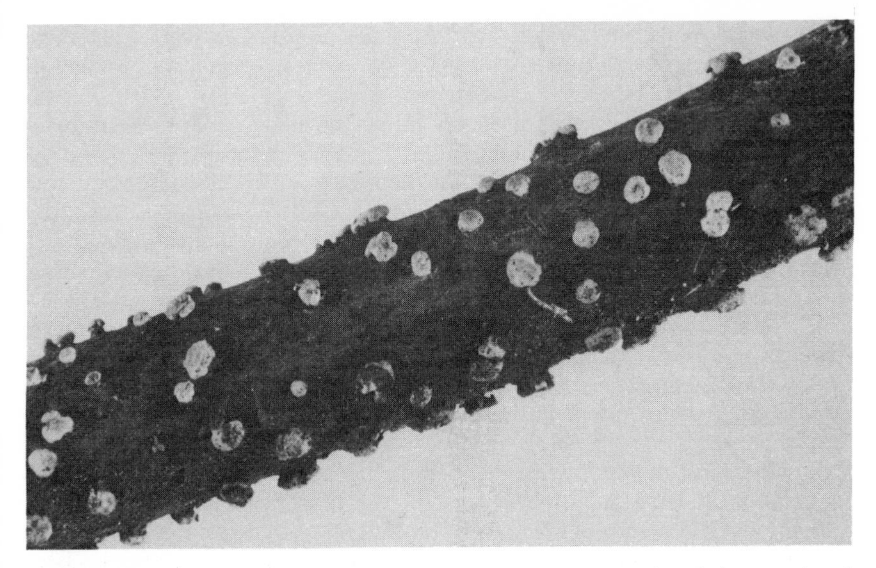

Fig. 5.1 *Nectria cinnabarina* growing on and in the bark of a twig and producing pustules of spores through the bark surface: this disease is known as coral spot.

The fungus may form a discrete lesion or spread to girdle and kill the tree. *C. corticale* seems a rather weak pathogen and may be associated with squirrel damage or possibly other sources of bark damage or general poor health of the tree caused by drought in especially hot summers.

The pathogens noted above are all attacking damaged or previously weakened trees. Another fungus, *Endothia parasitica*, colonizes sweet chestnut bark. It is thought that it was endemic (normally present at low levels) in China but it was introduced to New York in 1904 and the American chestnut had no resistance to it. All large chestnuts in the eastern seaboard were killed in a few years, though they continued to throw stool shoots which became infected when only a few inches in diameter. In Asia the fungus affects weakened trees, but in America it is a virulent pathogen because the host has no resistance. This is frequently the case with introduced pathogens and has been demonstrated yet again with Dutch elm disease (*Ceratocystis ulm*), a virulent strain of which has been introduced into Britain wiping out the elms which had lived for many years in some sort of rough balance with the indigenous strain of the fungus. *Endothia* has now been introduced into Europe and is affecting some chestnut plantations especially of the American species, *Castanea dentata*. The European species, *C. sativa* may be slightly resistant to the disease. These examples of introduced pathogens causing widespread damage are the reason for the rigid plant quarantine laws, which most countries have, to try to keep out non-indigenous strains or species of pathogen.

When considering the surface bark flora there is really very little information. There are scattered references to particular trees, usually in temperate forests, but there seems to be no concensus as to whether each tree species has its own heterotrophic flora. Since most of the nutrients for growth come from outside the tree, it is likely that, apart from the importance of bark texture, the flora is governed by nutrient availability, climate, degree of shading or exposure, etc.

The algal flora of trees is probably limited by moisture availability but it contains many different species (Round, 1981). The most profuse development is in rain forests, both tropical and temperate, and in cloud forests. Green algae such as *Chlamydomonas*, *Chlorococcum*, *Chlorella* and *Trentepohlia* are very common and non-gymnosperms also have some diatoms and cyanobacteria. There seems to be some evidence for different algal flora associated with differented trees though this is by no means certain.

Lichens have been studied very much more than the algae which are one of the symbiotic components. Lichens may form a significant part of the biomass in favourable conditions, such as the coniferous rain forest of the pacific coast of North America. Here just one species (*Lecidea erratica*) on Douglas fir appeared on three-year-old bark and could cover up to 25% of the surface of twigs from five to twenty years old. In the upper canopy frondose lichens in general could give 50 to 60% cover of the branches and twigs. In these conditions the leaching of organic carbon, and nitrogen from those species with cyanobacterial symbionts, may provide a significant amount of the available nutrients for other epiphytes. This production of available nitrogen by lichens is also true of tropical forests.

There are very extensive studies of lichens in temperate and boreal forests (Seaward, 1973). Here again they may be important in biomass terms, with figures of up to 500 kg ha^{-1} being common. There are distinct communities on each tree species and at different aspects and heights up the trunk. The surrounding forest type also influences the community present by affecting humidity, illumination, continuity of growth, air pollution, etc. These influences on communities (called alliances) have been well worked out for British forests and are illustrated in Fig. 5.2. It should be stressed that these are not species changes, but changes in whole communities of species brought about by the factors indicated. The eleven alliances included in the diagram have been carefully characterized as regards their habitat and species composition.

Lichens have been especially studied in relation to pollution, mainly by sulphur dioxide though other substances are also involved. A few lichens (e.g. *Lecanora conizaeoides*, in the *Leconorion variae* alliance in Fig. 5.2) are sulphur dioxide resistant and have become very common in industrialized areas where mean winter sulphur dioxide levels are 55–150 $\mu g\ m^{-3}$. The more usual response is however a reduction in lichen growth and in species diversity under sulphur dioxide stress. It is possible to use corticolous lichens as indicators of pollution levels, though other factors illustrated in Fig. 5.2 must of course also be considered. Considering two bark types in England and Wales the communities of lichens have been defined in relation to sulphur dioxide levels (Table 5.1): an alga or algal assemblage, *Pleurococcus viridis*, is also included in this table. The precise species may vary in different parts of the world but the principle will hold.

The fungi on bark surfaces are also very varied but have been studied much less systematically than the lichens. In general they are imperfects

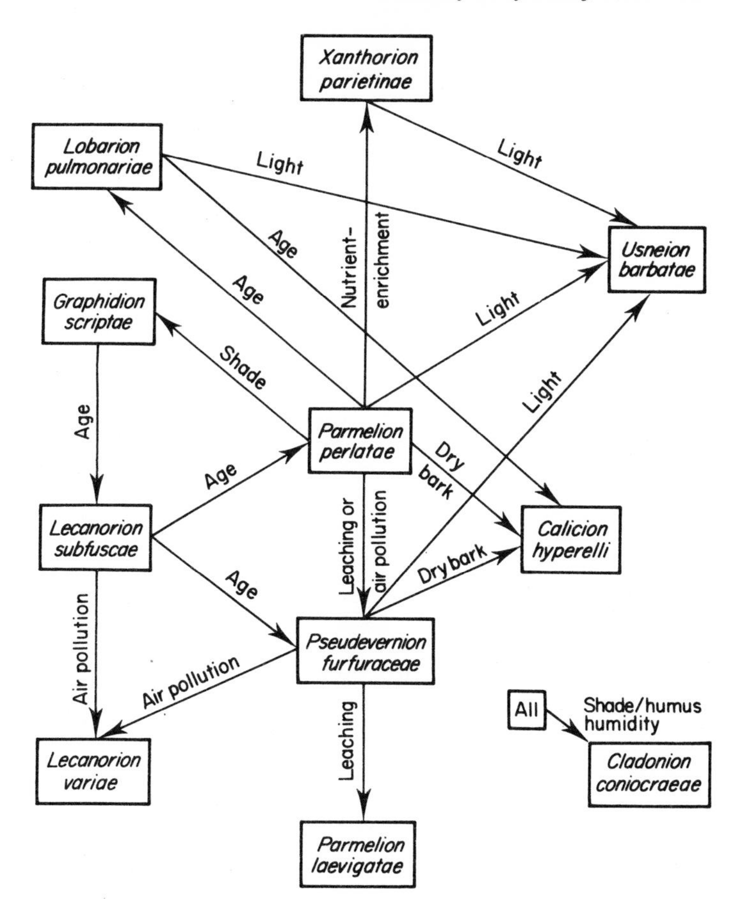

Fig. 5.2 Principal relationships between the epiphytic lichen alliances present in the British Isles. The lichen communities indicated are affected by factors shown on the arrows so that the community changes to a new composition. (From James, P.W., Hawkesworth, D.L. & Rose, F. in Seaward, 1973.)

(Deuteromycotina) and ascomycetes, with occasional isolates of zygomycetes and basidiomycetes. Many myxomycetes (*Arcyria*, *Comatricha*) are known to fruit on bark, and are active in the vegetative form when the bark is wet. Since there have been so few studies we may take the examples of fungal communities for each host tree.

Douglas fir has extensive bark flora when growing in the wetter parts of its natural range (Carroll *et al.*, 1980) and there is a changing flora as the twig surface ages, the maximum biomass (12–20 kg ha^{-1}) occurring at six or seven years. The fungi, which account for 35–60% of biovolume on twigs, are dominated by bark-inhabiting loculoascomycetes such as *Winteria caerulea* on younger twigs and *Massarina* sp. on older ones, and by sooty moulds in the Capnodiaceae (*Metacapnodium*, *Anntenulata*). The ubiquitous *Cladosporium*,

Table 5.1 Zone scale for the estimation of mean winter sulphur dioxide levels in England and Wales using co icolous lichens. (From Hawksworth, D.L. and Rose, F., 1976, *Lichens as pollution monitors*. Edward Arnold, London. Studies in Biology no. 66. pp. 64.)

Zone	Moderately acid bark	Basic or nutrient-enriched bark	Mean winter SO_2 ($\mu g\ m^{-3}$)
0	Epiphytes absent	Epiphytes absent	?
1	*Pleurococcus viridis* s.l. present but confined to the base	*Pleurococcus viridis* s.l. extends up the trunk	> 170
2	*Pleurococcus viridis* s.l. extends up the trunk; *Lecanora conizaeoides* present but confined to the bases	*Lecanora conizaeoides* abundant; *L. expallens* occurs occasionally on the bases	About 150
3	*Lecanora conizaeoides* extends up the trunk; *Lepraria incana* becomes frequent on the bases	*Lecanora expallens* and *Buellia punctata* abundant; *B. canescens* appears	About 125
4	*Hypogymnia physodes* and/or *Parmelia saxatilis*, or *P. sulcata* appear on the bases but do not extend up the trunks. *Lecidea scalaris, Lecanora expallens* and *Chaenotheca ferruginea* often present	*Buellia canescens* common; *Physcia adsendens* and *Xanthoria parietina* appear on the bases; *Physcia tribacia* appear in S	About 70
5	*Hypogymnia physodes* or *P. saxatalis* extends up the trunk *to 2.5 m or more; P. glabratula, P. subrudecta, Parmeliopsis ambigua* and *Lecanora chlarotera* appear; *Calicium viride, Lepraria candelaris* and *Pertusaria amara* may occur; *Ramalina farinacea* and *Evernia prunastri* if present largely confined to the bases; *Platismatia glauca* may be present on horizontal branches	*Physconia grisea, P. farrea, Buellia alboatra, Physcia orbicularis, P. tenella, Ramalina farinacea, Haematomma ochroleucum* var. *porphyrium, Schismatomma decolorans, Xanthoria candelaria, Opegrapha varia* and *O. vulgata* appear; *Buellia canescens* and *X. parietina* common; *Parmelia acetabulum* appear in E	About 60
6	*P. caperata* present at least on the base; rich in species of *Pertusaria* (e.g., *P. albescens, P. hymenea*) and *Parmelia* (e.g., *P. revoluta* (except in NE), *P. tiliacea, P. exasperatula* (in N)); *Graphis elegans* appearing; *Pseudevernia furfuracea* and *Alectoria fuscescens* present in upland areas	*Pertusaria albescens, Physconia pulverulenta, Physciopsis adglutinata, Arthropyrenia gemmata, Caloplaca luteoalba, Xanthoria polycarpa* and *Lecania cyrtella* appear; *Physconia grisea, Physcia orbicularis, Opegrapha varia* and *O. vulgata* became abundant	About 50
7	*Parmelia caperata, P. revoluta* (except in NE), *P. tiliacea, P. exasperatula* (in N) extend up the trunk; *Usnea subflori-dana, Pertusaria hemisphaerica, Rinodina roboris* (in S) and *Arthonia impolita* (in E) appear	*Physcia aipolia, Anaptychia ciliaris, Bacidia rubella, Ramalina fastigiata, Candelaria concolor* and *Arthopyrenia biformis* appear	About 40
8	*Usnea ceratina, Parmelia perlata* or *P. reticulata* (S and W) appear; *Rinodina roboris* extends up the trunk (in S); *Norman-dina pulchella* and *U. rubiginea* (in S) usually present	*Physcia aipolia* abundant; *Anaptychia ciliaris* occurs in fruit; *Parmelia perlata, P. reticulata* (in S and W), *Gyalecta flotowii, Ramalina obtusata, R. pollinaria* and *Desmazieria ever-nioides* appear	About 35
9	*Lobaria pulmonaria, L. amplissima, Pachyphiale cornea, Dimerella lutea*, or *Usnea florida* present; if these absent crustose flora well developed with often more than 25 species on larger well lit trees	*Ramalina calicaris, R. fraxinea, R. subfarinacea, Physcia leptalea, Caloplaca aurantiaca* and *C. cerina* appear	< 30
10	*L. amplissima, L. scrobiculata, Sticta limbata, Pannaria* spp., *Usnea articulata, U. filipendulla* or *Teloschistes flavicans* present to locally abundant	As 9	'Pure'

Epicoccum and *Aureobasidium* also occur. Resin deposits on Douglas fir have their own special fungi such as *Strigopodia batistae*.

Populus tremuloides (quaking aspen) has been studied particularly in relation to *Hypoxylon* canker. Numerous saprotrophic fungi have been isolated including many common deuteromycetes and some ascomycetes. Again there is suggestion that the community may vary with plant age. The saprotrophs can affect the pathogen in a variety of ways. Some of the yeasts and deuteromycetes inhibit *Hypoxylon* and so may reduce the instance of canker, and the saprotrophs may be stimulated by those bark extracts that inhibit *Hypoxylon*. Alternatively the saprotrophs may degrade pyrocatechol, so removing pathogen inhibition.

There have been similar investigations of beech bark flora (Cotter & Blanchard, 1982); of the isolations made two thirds were deuteromycetes, mostly Moniliales and a few genera accounted for a large proportion of the total: more than 25% of the fungi were *Alternaria; Aureobasidium, Cladosporium, Trichoderma* and *Epicoccum* were also common. Yeasts were approximately 10% of the fungi but zygomycetes, ascomycetes and basidiomycetes were rarely isolated. This particular study also recorded that bacteria were present on 24–69% of samples depending on the location of the trees. Protozoa (unidentified, but probably myxomycetes) were recorded from one fifth of the bark samples, being especially found in the sites of the highest bacterial numbers (Table 5.2 shows a selection of data from these studies on beech bark flora).

Actual numbers of organisms, or biomass, are rarely recorded but the results of one study of peach bark show great variation from sample to sample (Table 5.3). The quite low numbers, compared with leaf surfaces or soil for example, probably indicate a very patchy distribution with micro-colonies widely spaced so that they are missed altogether on some occasions. Bacteria were the least

Table 5.2 Micro-organisms isolated from non-cankered bark of American beech in New Hampshire. (Modified from Cotter, H. van T. & Blanchard, R.O., 1982.)

Micro-organisms	Isolation frequency (%)
Number of bark chips sampled: 96	
Algae	30
Bacteria	47
Protozoa	19
Fungi	214
Number of genera and form-genera represented: 22	
Alternaria sp.	6
Cladosporium sp.	22
Fusarium sp.	20
Penicillium sp.	13
Rhinocladiella sp.	7
Trichoderma sp.	7
Coniothyrium sp.	14
Cytospora sp.	6
Phoma sp.	22
Yeasts	13

Table 5.3 Numbers of bacteria and fungi per gram of peach bark. (Selected from Wensley, R.N., 1971, *Canadian Journal of Microbiology,* **17**, 333–7.)

	Date	Peach variety	Bacteria	Fungi
Bark surface sterilized	April	1	1500	0
		2	7000	250
		3	750	0
		4	250	500
	August	1	3750	0
		2	2750	0
		3	7250	0
		4	4350	0
	October	1	1250	250
		2	1750	250
		3	500	0
		4	1750	0
Unsterilized	January	1	2250	25 250
		2	6500	31 750
		3	1500	9250
		4	1500	34 000

variable components of the microflora, and many were again antagonistic to canker-forming fungi. It is suggested that the different degrees of resistance to canker displayed by the different varieties of peach may be related to the bark flora. Varieties 2 and 4 for example often have more saprophytic fungi and bacteria than the others. Notice how few of the microbes actually live within the bark: most are surface dwellers, especially the fungi. The very high numbers in non-sterilized bark may well be mostly surface spores which are not actively growing.

These few studies suggest that there is a common, fairly consistent flora of non-tropical trees. There may be some species or generic differences, usually related to bark characteristics, but there are also ubiquitous fungi common on all those trees examined. There is no information available from the tropical bark epiphytes except for the algae (see above). There must be a lot of fungi and bacteria in the moist conditions of tropical rain forests but no-one has so far looked at them. It would be prudent to hurry with such investigations, for uncut climax rain forest is rapidly becoming a rare community.

There is a specialized bark flora in the presence of slime fluxes, which are sap-flows, usually of pathological origin, from wounds or cracks in the stem. These are unusual habitats for bark organisms, being permanently wet and rich in nutrients. They are almost exclusively colonized by yeasts such as *Pichia pastoris* and *Trichosporon penicillatum* amongst many others: *Debaryomyces fluxorum* occurs in exudates from many species of tree. Numbers of yeasts often reach a peak at spring sap rise. There have been the usual problems with the identification of isolates, and modern analysis methods such as cluster analysis and principal component analysis have tended to group the yeasts by morphological and physiological characters such as sugar utilization, optimum temperature for growth, etc., rather than identify them to particular taxa. In an extensive survey of the literature, and new work in the field, Lachance, Metcalf and Starmer have looked at yeasts isolated from many localities from fluxes from different tree species. *Pichia*, *Candida* and *Hansenula* were most common but there was variation with the geographical distribution of the host and some indication of differences between host trees in their flora. When physiologically-determined groups, rather than particular taxa, were examined they were definitely related to the angiosperm host tree. Figure 5.3 shows a principal component analysis of some of the data; the two components Y1 and Y2 are highly correlated with sugar utilization by the yeasts (Table 5.4). The yeasts associated with each tree genus are therefore separated on the

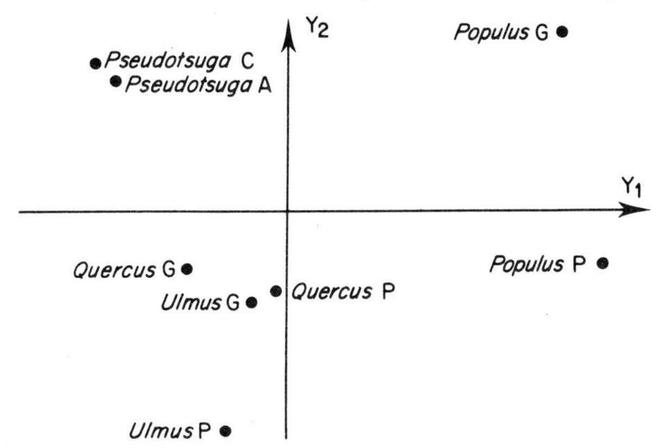

Fig. 5.3 Ordination of the physiological profiles of eight yeast communities based on two principal components, Y1 (50.6% of the variation) and Y2 (24.2% of the variation). The geographic origins are indicated by A = Arizona, C = California, G = Great Lakes region and P = Pacific North-West region. (From Lachance, M.A., Metcalf, B.J. & Starmer, W.T., 1982, *Microbial Ecology*, **8**, 191–8.)

Table 5.4 Physiological traits significantly correlated with co-ordinates of principal components Y1 and Y2 from Fig. 5.3. (From Lachance, M.A. *et al.*, see Fig. 5.3.)

Traits	Confidence levels	Correlation with Y1	Correlation with Y2
Growth on cellobiose	99.9%	0.96	
Growth on salicin		0.94	
Growth on rhamnose		0.94	
Growth on glycerol	99%	0.90	
Growth on trehalose		0.88	
Fermentation of glucose		0.87	
Growth on erythritol		−0.84	
Growth on ribose	95%		0.80
Growth on succinate			−0.80
Growth on galactose			−0.79
Growth on inulin			−0.75
Growth on galacititol			−0.74
Growth on melibiose			−0.73
Growth on raffinose			−0.73
Growth on D-arabinose			−0.72

basis of their sugar utilization so that the gymnosperm, *Pseudotsuga*, obviously has a different flora from the angiosperms. Thus the yeast population of *Populus* is different from *Quercus* and *Ulmus* which are, however, quite similar to each other. This presumably reflects differences or similarities in the chemical composition of the sap from different genera. *Populus* also showed a geographical effect, yeasts isolated from poplars in the Great Lakes region being different from those in the Pacific North-West of America. When relationships amongst the host trees were also considered in terms of chemical taxonomy further analysis showed good correlation between yeast groups and tree groups.

Decay of woody stems

General considerations

This is a subject on which there is a vast literature because of its economic importance, and it includes timber decay during conversion, storage and use (see Liese, 1975; Levy, 1982). There are extensive lists of organisms known to cause decay, mostly basidiomycetes with lesser numbers of ascomycetes and some bacteria. These are often based on the large, obvious fruiting bodies produced by the basidiomycetes which may not reflect the importance of the mycelial phase, which is causing the decay. The basic chemistry of timber decay will be considered, but the colonization and microbial interactions in the standing tree are the main subjects of this section.

Wood is a complex substrate, both chemically and spatially. Within the tree there is phloem on the outside of the cambium and this often becomes mixed with the bark and may be shed with bark flakes. The xylem in the centre of the tree is composed of various cell types, mainly tracheids and parenchyma and also vessels in angiosperms. The open cell lumen, of vessels especially, may allow colonization without having to penetrate walls. The few annual rings next to the cambium are the sapwood and these have live parenchyma with protoplasts and lipid and starch as storage materials. Vessels and tracheids are dead and devoid of cytoplasm after they have differentiated. Nearer the centre of the tree, in the heartwood, all cells are dead and without storage materials; there may be additional deposits of phenols and other 'extractives' in the heartwood. The proportion of the stem which is heartwood and the toxicity of the extractives varies from species to species. There are therefore two sorts of nutrients in wood, the readily available but transient materials in the sapwood and the more difficult to degrade polymers of the cell walls in both sapwood and heartwood. Because the sapwood is alive it has various defence mechanisms (p. 23) and in standing trees heartwood tends to be decayed while sapwood is resistant, hence hollow trees or trees with heart rots occur. In felled timber, in use by man, the reverse is true; sapwood though now dead, contains more easily available nutrients and is colonized first and the heartwood is more resistant.

Chemically the cells walls have three main components, cellulose, lignin and hemicellulose which may be utilized by the same or different micro-organisms to different extents. It is rare for a single organism to degrade the entire wall;

wood decay in natural conditions is concerned with successions of organisms and synergistic and antagonistic interactions.

In the living xylem of the sapwood the tracheids and vessels are of course filled with liquid. In the heartwood they contain air unless plugged with resin or tyloses. For the growth of most decay fungi an initial water content of about 25% is needed although higher levels increase growth up to about 40% on a wet weight basis. Above this the wood is waterlogged and becomes anaerobic, because of the low solubility of oxygen in water, and bacteria then predominate (p. 92). Different water contents affect the microbial species which can degrade the timber.

Levels of mineral nutrients in wood are usually low in gravimetric terms (g g^{-1}), and attention has focused on nitrogen as a rate limiting factor, though phosphorus may also be in short supply. The actual amount of nitrogen in wood is however quite large on a weight per volume basis (2–3 g l^{-1}) especially when compared with levels in soil or oligotrophic aquatic environments. However, wood has a rather low density, it is mostly air with the cell walls and tracheids, etc., being the only solids, so when wood is wet the nitrogen is spread through the volume; in dry wood nutrients are concentrated on and in the walls of the wood cells. There may also be other nitrogen sources; nitrogen-fixing bacteria occur in some decay situations and nitrogen is brought into the system in fungal spores and in arthropod bodies. Fungi move nitrogen about in the wood, concentrating it at the growing tips of their hyphae and re-using that from old, dead hyphae. There is tight nutrient cycling and nitrogen immobilization while the carbohydrate is respired.

The C:N ratio is 350:1 to 500:1 or even higher, and it varies from sapwood to heartwood. As decay proceeds the loss of carbon as carbon dioxide reduces the C:N ratio until at about 25:1 nitrogen is not limiting growth, and it is released. Considerations of C:N ratios are however complicated because it is not really the total carbon and nitrogen which are controlling microbial growth, but the available carbon and nitrogen. the nitrogen may well be present in the chitin of fungal walls and so not generally available. For example in a study of *Castanea* decay there was a 39% weight loss in fifteen weeks, but 58% of what was left was fungal material containing nitrogen as chitin, so the decay of the wood had been much more rapid than the overall weight loss would suggest. Similarly much of the carbon is certainly not available because of the problems in the decay of lignin and cellulose. Soluble carbon with C:N ratio of 350:1 would certainly give nitrogen limitation but if the carbon is insoluble then the low nitrogen may be adequate or even super-optimal and thus the enormous amount of carbon is irrelevant because it is not available. Whether nitrogen is rate limiting depends not on the total C:N ratio, but on the ratio of available carbon to available nitrogen which is much harder to determine. The levels of carbon and nitrogen may influence the species present and some wood decay fungi seem especially suited to utilizing carbon at low available nitrogen.

Organisms causing decay

The organisms living in wood have been grouped both taxonomically and according to their position in supposed or actual successions (Käärik, 1974; Levy, 1982). It is important to know what kind of organisms are present, for not all can degrade cell wall polymers. Bacteria are mainly Gram-positive rods (*Bacillus, Clostridium*) but *Pseudomonas* and *Corynebacterium* also occur, and there are actinomycetes in wood. As noted above some of these bacteria may fix nitrogen, and if nitrogen is limiting this could be important, and such nitrogen could be available to other organisms, especially the important wood rotting fungi. Bacteria can be synergistic in other ways too; they break down pit membranes, particularly in ray parenchyma in sapwood and this increases the aeration of the timber, allows water access and also the passage of organisms which cannot themselves break down wood. Conversely some bacteria are antagonistic to decay fungi.

There are *bacteria* that break down the main cell wall components but in general they are not important in direct decay except in certain specialized situations, usually when the wall has a very high moisture content. This may occur in the storage of freshly-felled timber in log ponds prior to processing. These storage problems are not to be confused with the occurrence of bacteria in standing trees, especially poplars, in a pathological condition called wetwood. Water-soaked, anaerobic xylem occurs in this disease and it presents problems after sawing because it bends, checks and cracks on drying. Pit membranes are destroyed, but the actual amount of decay of timber is not significant. *Erwinia, Xanthomonas, Acinetobacter, Clostridium, Bacteroides*, sulphur-reducing and methane-producing bacteria can be isolated in quite high numbers. Normal sapwood contains $10^2 - 10^3$ facultative anaerobic heterotrophs per g, less than 10 methanogens and 10^2 nitrogen-fixers: wetwood has 10^6 to 10^7 facultative anaerobes, $10^3 - 10^4$ methanogens and $10^4 - 10^6$ nitrogen-fixers.

Primary moulds, which are '*r*' strategists and not capable of degrading the wood, invade rays and the end grain and utilize starch, lipids, etc., in the sapwood parenchyma. They are mostly deuteromycetes, or their ascomycete sexual stages. *Staining fungi* (*Alternaria, Phialophora*) occupy the same cells, though some of them can penetrate the walls, and they have pigmented hyphae and cause a blue or greenish stain especially in sawn conifer timber. Many of the symbionts carried by bark beetles and other lignicolous insects (Batra, 1979) are staining fungi.

Secondary moulds (*Trichoderma, Gliocladium*) occur after the main decay period: again they do not usually utilize lignin and cellulose but they use the by-products of other organisms, such as the dead hyphae or soluble carbon released by extracellular enzymes of true wood decomposers. The role of *protozoa* (including myxomycete fungi) in wood decay had not been well determined. Table 5.2 shows protozoa in bark, myxomycetes have also been isolated from within both healthy and diseased wood. They must depend on pre-existing holes to gain entry and probably live, as amoebae, on the bacteria that are present. It is unlikely that the myxomycetes are doing much actual decay.

The main *timber decay fungi* are ascomycetes and basidiomycetes. Ascomycetes and their imperfect stages cause soft rots, usually in converted timber. There are many genera reported but common ones are *Chaetomium, Xylaria, Peziza, Alternaria* and *Phialophora*. The hyphae occur in the ray parenchyma, then in the secondary walls (S_2 layer) of the tracheid or vessel,

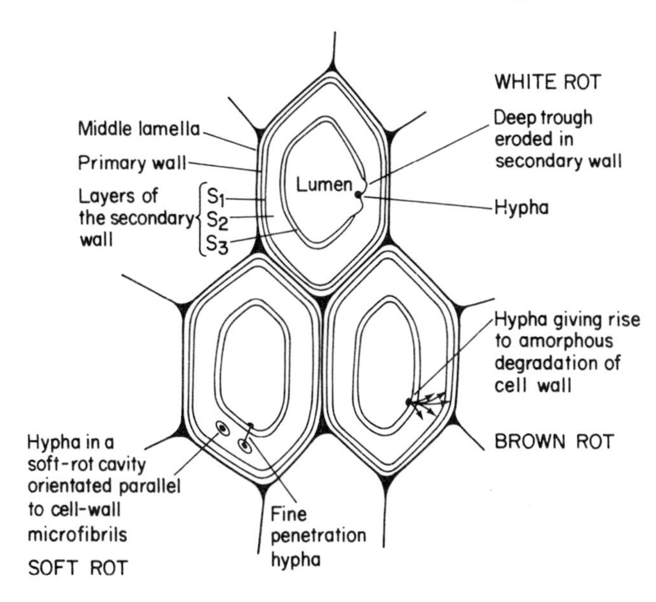

Fig. 5.4 Diagram of wood cells in transverse section showing patterns of degradation produced by three types of wood decay. (From Montgomery, 1982, in Frankland *et al.*, 1982.)

forming cavities with pointed ends which often run round the cell following the alignment of the cellulose fibrils. In transverse section these appear as holes in the wall with an hypha in the centre of each (Fig. 5.4). Eventually the holes enlarge and meet to form a large cavity. In hardwoods there is also some growth within the lumen. Cellulose and hemicellulose in the wall are attacked though not usually the lignin to any extent. The weight loss is often quite slow (25% in three months) though it does vary with the species of timber. Soft rots are important when the wood is wet or when basidiomycetes have been eliminated, by biocides for example. Small sizes of timber are particularly susceptible, for soft rots only grow in the surface layers. This may not matter in large sizes of timber such as are used in the construction of piers, docks and jetties, though abrasion may be a serious problem. However in small, especially softwood, slats such as are used in water-cooling towers and in small boat timbers the effect may be serious, with complete destruction and loss of strength of the timber. Prevention of soft rot attack cannot usually depend on keeping the timber dry, and phenolic preservatives such as creosote are not effective. Usually heavy metal salts (copper or chromium) and arsenic are used in various complex formulations to control the solubility, wettability and penetration of the biocides.

Basidiomycete rots have received by far the most attention. The hyphae of basidiomycetes generally occur in the cell lumen or pressed onto the wall surface, not actually growing within the wall. They do penetrate and cross from cell to cell by narrow bore holes going straight across the walls. There are basically two types of basidiomycete rots, brown and white. Brown rot fungi decay cellulose and hemicellulose; the lignin is left more or less unaltered and the wood turns brown. Since the lignin is left the walls retain their structural integrity until a late stage in the decay and macroscopically may look almost normal, though the timber has lost a lot of its strength (Fig. 5.4). The attack is mainly on the S_2 and S_1 walls, the primary wall and the middle lamella are not affected.

Basidiomycete white rots can degrade lignin, and other phenolic compounds as well as the polysaccharide, though all components are not degraded to the same extent by different organisms. The whole cell wall substance is therefore removed, including the middle lamella in some cases. The walls close to the hyphae become visibly thinner so that each hypha is left on a ridge of wall material with eroded areas on each side (Fig. 5.4).

Chemistry of decay

The chemistry of decay has been well studied, but many aspects are still not understood. The long cellulose molecules occur in crystalline or para-crystalline areas called micelles, with one molecule passing through several different micelles. Lignin and hemicelluloses and water, fill spaces between the cellulose molecules. Cellulose is a polymer of D-glucose mostly linked β, 1–4 and is hydrolysed by cellulases. There are several different components in most cellulases which have been given many names over the years, but basically there is an exoglucanase which splits off the dimer cellobiose or the monomer glucose from the non-reducing ends of the molecule. This enzyme does not operate on crystalline cellulose. Secondly there are several endoglucanases which can disrupt the cellulose molecule at many points, giving random lengths of glucose polymer. The endoglucanases can break up crystalline cellulose and produce lots of chain ends available for the action of the exoglucanase. There are also 1–4, β-glucosidases to hydrolyse cellobiose and small oligomers to glucose. Organisms vary in the number of enzymes or iso-enzymes they possess: as the technology of protein separation has improved the apparently simple 'cellulase' has been resolved into a complex mixture of related enzymes. Though hydrolysis is undoubtedly the main pathway there are various others known such as the oxidation of cellobiose.

Hemicelluloses are a mixed group of polymers of 5- and 6-carbon sugars, such as xylans and mannans. Most of the degradation is again hydrolysis, usually by endo-enzymes.

The lignins are a group of closely-related polymers of phenylpropane (Fig. 5.5). The gross structure, as well as the side chains, varies from species to species and between gymnosperms and angiosperms, though the complete structure of lignins is not known. Soft rot and brown rot fungi do not remove much lignin but they may alter side chains, especially methoxyl groups, and there is some limited ring cleavage. There is more demethylation by brown rots and these also produce polyphenoloxidase which may oxidize catechol units to quinoid or melanin pigments to give the brown colour.

White rot fungi are the most important lignin degraders. Side chains are shortened and oxidized and ether linkages joining the phenolic units together are broken, releasing ortho-diphenols. These are further degraded by ring cleavage and side chain oxidation. There is some evidence that bacteria,

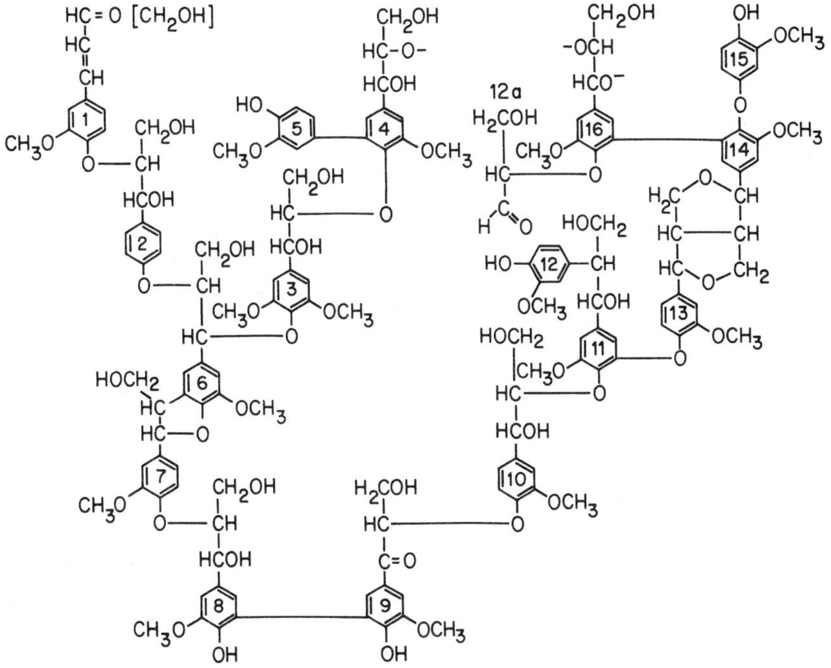

Fig. 5.5 Schematic formula for a part of spruce liginin consisting of 16 phenylpropane units. (From Ander & Eriksson, 1978; see also Adler, E., 1968, *Svensk Kemisk Tidskrift*, **80**, 279.)

although not important in the decay of lignin in the wood, may produce enzymes which degrade some of the initial products of lignin degradation by the white rot fungi. This is an example of the synergism mentioned above (p. 91). There is now a lot of information on the enzymes involved in these various reactions, including demethylases, oxygenases, dioxygenases, and phenol oxidases (Ander & Eriksson, 1978). Different white rot fungi remove different proportions of cell wall components. *Coriolus versicolor* removes lignin, cellulose and hemicellulose simultaneously but *Fomes ulmarius* uses the lignin faster and *Trametes pini* can delignify cell walls without affecting the cellulose.

Successions in timber

Because of the complexity of the structure of wood and the large numbers of organisms involved in its degradation it is often difficult to decide whether the organisms are pathogens or saprotrophs. Fungi attacking heartwood are not using live material, even though it is an integral part of the plant and the attack damages the plant: are they pathogens? Those attacking the sapwood are clearly necrotrophic pathogens. A great deal of timber decay starts in dead timber and then later spreads to the living stem. Most wood decay organisms need to gain access through a natural or artificial wound, though a few like *Armillaria* can penetrate live roots and subsequently spread up to kill an unwounded stem.

When woody stems are wounded there is a host response, irrespective of any microbial colonization. The wood becomes discoloured, usually brown because of phenol oxidation, and this reaction is intensified by subsequent microbial invasion. Vessels, in hardwoods, become plugged above and below the wound; rays limit spread laterally and annual rings delimit the discoloured area radially (see Fig. 5.7). Thus a column of discoloured wood is formed, stretching rather more above than below the wound, which may inhibit microbial invasion by phenol deposition in angiosperms or resin in conifers. Conifers may form special traumatic resin canals.

The early colonizers which are bacteria, primary moulds and staining fungi (p. 92) may be capable of degrading the phenols in discoloured wood, and this could prepare the infection court for invasion by phenol-sensitive decay basidiomycetes. *Chondrostereum purpureum* and *Bjerkandera adusta* however are unaffected by phenols in discoloured wood. Pioneer primary fungi may also be antagonistic to the basidiomycetes by (1) nutrient competition including nitrogen immobilization (e.g. by *Hypoxylon multiforme*) or (2) by the production of antibiotics by fungi such as *Scytalidium*. Actinomycetes such as *Streptomyces parvullus* may also use phenols and/or inhibit basidiomycetes.

These sequences of organisms are affected by many environmental factors, especially nutrients, moisture content and temperature. An extreme case of this is the decay of dead branch wood (see also 100 – 102) still attached to the tree, for it is very dry for much of the time and the decay succession is therefore extended. In a study of poplar, bacteria and micro-fungi (e.g. *Cystospora*) invaded and persisted to attain dominance after four years. Later micro-fungi such as *Phoma* were dominant at six years but were themselves replaced by basidiomycetes like *Bjerkandera adusta* after eight or nine years. Some basidiomycetes (e.g. *Phellinus* and *Fomes igniarius*) only occurred in branches which had been dead for more than nineteen years. Decay may proceed much more rapidly if the branch should fall to the ground and become embedded in moist litter. This succession is therefore limited by moisture but changes in hydrogen ion concentration have also been suggested as a possible factor in some successions on Balsam fir associated with bark beetles: *Ophiostoma bicolor* and *Stereum chailletii* were replaced by *Polyporus abietinus* after one year. It is however likely that nutrients and moisture are usually the dominant factors. Successions may also vary with the position on the tree (Fig. 5.6). Inoculation of specialized fungi by bark beetles and other wood-boring insects can also modify the succession so that the initial stages are dominated by the insect symbionts which are later replaced by wood decay basidiomycetes (Fig. 5.6).

The above generalizations on succession do in fact mask a much more complex situation which has been the subject of detailed analysis. In order to study the community structure of wood decay basidiomycetes the interactions of mycelia must be studied rather than just observing the fruit bodies on the outside of the stem. Each mycelium may compete for space and nutrients with its neighbours and it occupies a unique niche, though not a unique micro-habitat for it will share the space with other secondary fungi and certainly with bacteria. Different mycelia can be distinguished by colour or textural

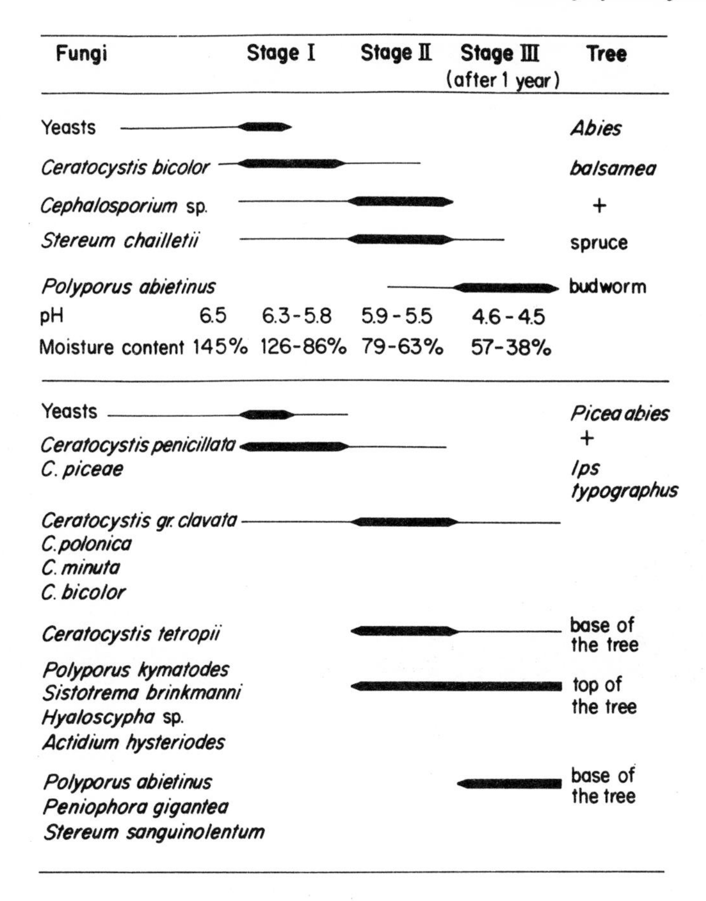

Fig. 5.6 Successions of fungi in insect-deteriorated standing trees. (From Käalik, 1975, in Liese, 1975.)

differences in the wood, reflecting different types or stages of decay, and mycelia are often delimited by zone lines. These are dark lines which separate rotten areas from sound ones or separate zones occupied by different rot-causing fungi (Figs 5.7 and 5.8). The dark colour may be due to tannins and to gums laid down by the tree (as for infection by *Ganoderma adspersum*) but more usually they are caused by interaction zones between different fungal mycelia. The observation of zone lines is supplemented by cultural studies of compatability and also by the identification of the isolates. In general the more distantly related the strains, the more intense the interaction between their mycelia.

The timber may be colonized by a single isolate of a single species (Fig. 5.7) so there is a single decay column. This may occur in the case of a pathogen causing heart rot where there was a single infection court. With most decay in woody stems, however, there is extensive colonization by different strains of the same species, and by several different species, so that the single term

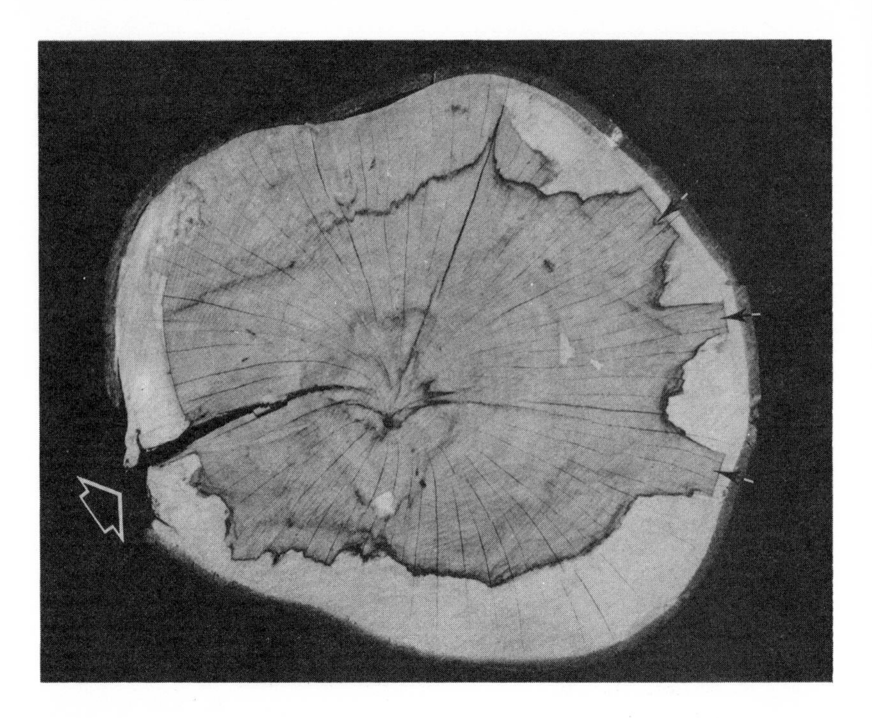

Fig. 5.7 Decay column caused by a single infection (at the wound, large arrowhead), with zone lines separating the infected timber from the healthy. Notice how the decay stops at an annual ring in 3 places (small arrows).

'basidiomycetes' in the successions described above does in fact cover a complex set of interactions and successions within it. For example, even though a stem may be colonized by just *Coriolus versicolor* it may produce many morphologically different fruiting bodies, varying in colour, zonation, size, etc., and each fruiting body type generally is associated with a particular decayed volume (a decay column) separated from others by zone lines (Fig. 5.8). Isolates from different decay columns do not grow together in culture so that as well as sexual incompatability systems there are vegetative mechanisms to delimit individuals of an interbreeding population. This is a complication with these studies, in that basidiomycetes have the very prolonged dikaryon stage in which two nuclei exist in each hyphal compartment. Normal, haploid basidiospores of heterothallic species germinate to give a haploid, primary, mycelium and compatible mycelia may fuse to give the dikaryon which eventually forms a fruit body, though this may take several years. In the fruit body karyogamy and meiosis occur. The fusion of haploid mycelia (e.g. of *C. versicolor* or *Bjerkandera adusta*) is governed by sexual compatibility factors. The merging or more usually the antagonism of the resulting dikaryons is governed by vegetative factors. Some homothallic species (*Stereum hirsutum*) have multinucleate spores that form primary mycelia that are multinucleate

Fig. 5.8 Population and community organization in a beech log which has been placed upright with its base buried in soil and litter two years previously at a woodland site. The log has been cut transversely near the top and bottom surfaces, and longitudinally for the intervening length, so that the three-dimensional distribution of decay and discolouration can be ascertained. The distribution of some of the decay fungi present is indicated by the following symbols: •, different individuals of *Coriolus versicolor* demarcated from each other by interaction zone lines, *C. versicolor* is fruiting on the top of the log. Ab, *Armillaria bulbosa* invading peripherally from the base. Cm, *Chaetosphaeria myriocarpa* (conidial stage) occupying the interaction zone and other regions not occupied by active decay fungi. Pv, *Phanerochaete velutina* invading from the base. Sh, *Stereum hirsutum*. (Photograph courtesy A.D.M. Rayner, University of Bath. Based on Coates, D., 1984, *The biology of intraspecific antagonism in wood decay fungi*. Ph.D. Thesis, University of Bath.)

and these form discrete decay columns and are antagonistic from the outset, even if the cultures are those from spores of the same fruiting body.

Interspecific interactions occur very commonly and if there are interactions which result in one species replacing another, then successions will occur. It is very difficult to study these in natural conditions since the sampling is destructive and successions imply changes in species composition or dominance over time in one habitat, requiring repeated sampling. However, culture studies or inoculation of small wood blocks may show the types of possible interactions *in vitro* and these may then be recognized in a natural resource and successions inferred. In culture, species may overgrow and mix, though this is unusual, they may overgrow and replace one another by lysis of the overgrown culture or there may be a deadlock situation where the cultures meet, a zone line forms and neither invades the other. There is a heirarchy of replacement; for example *Ganoderma adspersum* and *Laetiporus sulphureus* can be replaced by *Stereum hirsutum* which in turn yields to *Coriolus versicolor* and decay columns of this fungus can then be colonized by *Bjerkandera adusta* or *Hypholoma fasciculare*. Replacement may be due to nutrient competition or the exhaustion of nutrients for the first species and its replacement by another with different nutrient requirements which can be met by the depleted substrate. There are also direct hyphal interactions and the lysis of the initial colonizers in natural situations. Deadlock interactions occur between particular species; thus *Bjerkandera adusta* deadlocks *B. fumosa* and *Coryne sarcoides* (an ascomycete) and *C. versicolor* deadlocks *Armillaria mellea*, *Daedaleopsis confragosa*, *Ganoderma applanata* and *Phlebia merismoides* amongst others.

There is some evidence from more natural situations that these cultural findings are valid. Where stems are known to be infected with particular fungi they can be artificially inoculated with fungi known from the culture sequence to replace or deadlock the initial colonizer. Subsequent analysis is done to verify what in fact happened and this turns out to be remarkably similar to the predictions from culture.

Some of these interactions are illustrated in the development of a hypothetical community in Fig. 5.9 (see details in legend). This shows the problems in studying natural communities by sampling at a given time, and indicates that the succession described for timber (p. 95) should be viewed as simplified versions of what may be happening.

The colonization of branches of oak has also been studied by these detailed analysis methods. Many branches are colonized by a single species, of which *Stereum gausapatum* is the most common. This species and *Phlebia rufa* are pioneers and colonize living or recently dead branches. *C. versicolor* and *Phlebia radiata* secondary colonizers and may displace the pioneers. These studies are illustrated by the data for a single branch (Fig. 5.10) which shows just how complicated the decay patterns may be (see details in legend). The distribution of fungi in each cross-section of the branch is shown and this has been supplemented by culture work to check on the identity and compatibility of the different isolates. There is no evidence in this work of prior colonization by hyphomycetes (p. 96) and the discolouration of timber, which may reflect different methodology or the different species of tree. Also these oak branches were not obviously wounded either naturally or artificially.

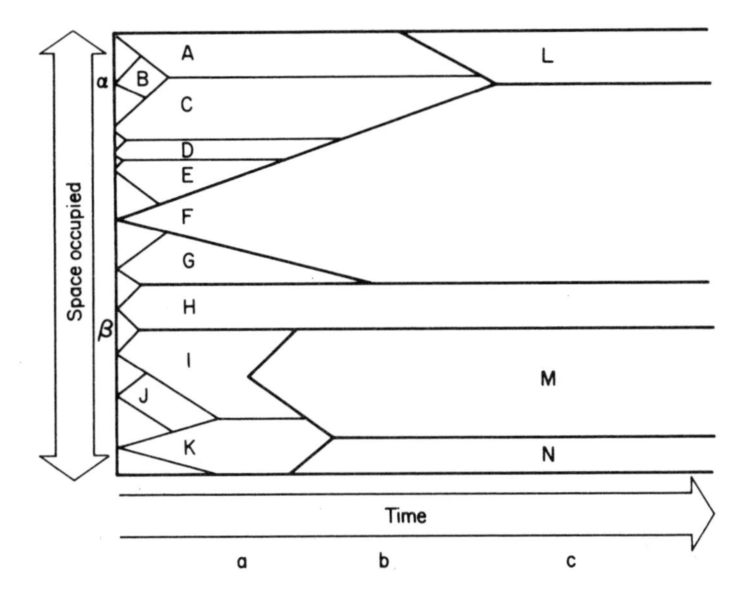

Fig. 5.9 Diagram illustrating the concept of fungal community development within a spatially defined resource. The total available colonizable space is represented by the gap between the top and bottom horizontal axes. The changing distribution, within this space, of a variety of different fungal mycelia A–N, over a period of time starting from the point at which the resource first becomes susceptible to fungal colonization (represented by the longitudinal axis at the left-hand edge of the diagram) is shown as a series of tapering and expanding bands. Horizontal lines between adjacent bands indicate deadlock interactions, whilst oblique ones indicate replacement. Community development in this hypothetical example can be summarized as follows. Mycelia A–K all colonize more or less from the outset and spread through the resource until eventually they all come into contact and the available space for colonization is filled. B and J occur for only a short period before being replaced. A and C, C and D, D and E, G and H, H and I and I and K are all involved in deadlock interactions, but F is able to replace C, D, E and G and becomes one of the dominants in the final community. Subsequently L, M, and N invade and replace other mycelia before becoming deadlocked, so that in the final community only L, F, H, M and N are present. The development of M in wood occupied initially by I could result from a specific mycoparastism, such as occurs in nature between *Pseudotrametes gibbosa* and *Bjerkandera adusta*. Successional interpretations of this complex pattern of events could easily lead to oversimplification and distortion. For example, at point α there is a 'succession' from B to A to L. Whilst this does tell us of the relative competitive abilities of these three mycelia, it does not tell us when and where they colonized and , of course provides only a narrow view of events in the whole resource. Even narrower would be the view at point β, where mycelium H persists throughout, no 'succession' in fact occurring. Alternatively, if the species assemblages present in the entire resource at different points in time a, b and c were analysed, a broader picture would be obtained, yielding some information about when different mycelia colonized, but revealing little detail about specific instances of replacement and deadlock. (From Rayner & Todd, 1979.)

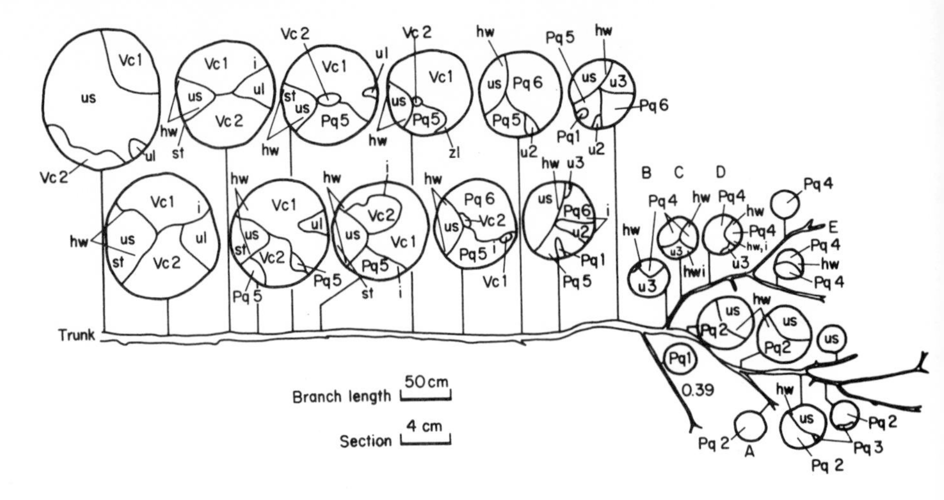

Fig. 5.10 Decay community structure in a branch of oak. Different basidiomycete individuals were demarcated by interactive zone lines (i) in the wood. The predominant species were *Vuillеminia comendens* (two individuals; Vc 1–2) and *Peniophora quercina* (six individuals; Pq 1–6). Three other unidentified basidiomycetes (U 1–3) occupied relatively smaller volumes. Both *V. comendens* and one of the *P. quercina* individuals occurred in the main stem possibly orginating from small branch stubs, or by direct entry. It is likely that Pq 2 originated in the branchlet from which section A was taken and progressed to the main fork whence it grew in both available directions. In the distal sections B–E, heartwood wings (hw) were present in the absence of living tissue (us) and separated decay columns occupied by either the same, or two different individuals. In the latter case, an interactive zone line was also present. Presumably the fungi were initially confined to regions between the heartwood wings, with living tissue outside. Subsequently, possibly following some traumatic event, the fungi were able, by some means, to penetrate or by-pass the wings, gaining access to the living wood and colonizing it without further wing formation. (From Boddy, L. & Rayner, A.D.M., 1983, Ecological roles of basidomycetes forming decay communties in attached oak branches. *New Phytologist,* **93**, 77–88.)

Diseases of woody stems

Many pathogens of woody stems have been mentioned in passing but we must consider a few examples of these briefly in their own right. Pathogens of woody stems are important in forestry because of the economic loss they cause, and wood is a major world resource.

Heterobasidion (Fomes) annusum is a most serious disease wherever conifers are grown in the world. In Britain it accounts for about 90% of the decay in standing conifers. Long range dispersal is by basidiospores which infect wounds or cut stumps after forest operations. The fungus grows into the timber and causes a rot of the roots but it also extends up the stem to 4 or 5 m, occasionally more, causing a heart rot. The rot starts as a small, white, pocket rot and finally becomes a white fibrous rot. Once it has infected a root system it spreads to neighbouring trees through natural root grafts, especially in single species plantations, and causes groups of dead or dying trees centred round an old stump (see p. 144, Chapter 6).

Polyporus schweinitzii is another serious wood-rotting organism that, though it rots roots, also causes serious decay in stems, especially in North America. It produces a brown powdery or cubical rot.

Armillaria mellea is again primarily a root invader, but it can be considered here for it kills stems and causes decay. The fungus is worldwide in distribution (both temperate and tropical) and very damaging on a forest scale and also to individual ornamental trees and shrubs. It is as important, or more so, as *Heterobasidion* because it attacks broad-leaved trees as well as conifers. This fungus is considered in more detail for its root rotting ability (p. 144).

There are hundreds of other pathogens of woody stems, mostly basidiomycetes, but also occasionally ascomycetes like *Rhizina undulata*, which occur in all forests. There are no control measures once infection occurs for two reasons: fungicides do not penetrate woody stems easily and the wood has a low value compared with agricultural crops and so it is not worth the expenditure of large sums of money on protectant or eradicant chemicals. Control is therefore by general forest hygiene and prevention of wounds and damage.

There are, however, some diseases of woody stems that can to some extent be cured. Silver leaf of fruit trees is caused by the fungus *Chondrostereum purpureum* which is a wood rot in the stems and branches, but also causes the leaf epidermis to separate from the mesophyll and the resulting air space gives the leaf a silvery appearance. Death of the tree results if there is no treatment. Fungicide applied to the outside of the tree has no effect, injection of the stem is also ineffective. Use has recently been made of a biocontrol agent, inoculating holes bored in the stem with *Trichoderma* on wood dowels. Protection and some recovery has been recorded. Like most successful cases of biological control with introduced antagonists this is an introduction into a specialized environment with only limited competition. Healthy stems have very few other fungi apart from the introduced one. It is very difficult to introduce biocontrol agents into complex environments such as soils (p. 150).

There are important vascular wilt diseases affecting woody stems, including oak wilt and Dutch elm disease caused by *Ceratocystis fagacearum* and *C. ulmi* respectively. Oak wilt is spread by various bark beetles which emerge from diseased trees and then feed on neighbouring healthy ones, infecting them with the fungus. Birds and squirrels may also spread the fungus, and transmission through root grafts also occurs. Spread is in general rather slow and the disease, very important locally in the U.S.A., is not of worldwide importance. Dutch elm disease, though closely related, is altogether different. It is spread specifically and very efficiently, by two bark beetles, *Scolytus scolytus* and *S. multistriatus*, which are infected on emergence from the brood galleries and again infect neighbouring trees by feeding on young shoots. Root graft spread also occurs. The disease causes wilting and yellowing of the leaves and death of the branch, or even of the whole tree in one growing season. The disease is endemic in Europe and North America and, though there have been occasional outbreaks, it was only recently that a very virulent strain of the fungus arose killing many elms in North America and it has now been introduced into Britain, where almost all the elms have been killed. Some regrowth of suckers has occurred, but these will probably not survive to mature trees. Control was attempted by removing and burning all infected trees but this failed to contain the outbreak. Annual high-pressure injection of fungicides, although expensive, prolonged the life of some valuable ornamentals, but it is not feasible either logistically or economically on a wide scale. Similarly control of the vector by insecticides failed. There is now no serious control problem in lowland Britain because there are almost no host trees. The search for resistant strains and possible controls does, however, continue in the hope that replanting can be done with something resembling elms. Related genera such as *Zelkova* and some supposedly resistant strains of *Ulmus* are being used as well as other common genera. There has been a great economic loss from the waste of timber and much greater amenity loss, for elms were a very valuable city tree and in Britain also most important in hedgerows. *Ceratocystis* does not decay the timber though dark staining occurs, but the death is so rapid that most timber was not used as much as it should have been because of the glut on the market.

Conclusions

The microbiology of stems quite naturally divides between the general flora of green herbaceous stems, which is similar to that of leaves, and the specialized basidiomycetes of large, woody perennial stems. Serious stem diseases of agricultural crops are rather rare but are of course a major concern. Stem diseases have been controlled by a biological means, especially the crown gall bacterium and the important butt rots like *Heterobasidion*, partly because of the difficulty of chemically treating the inside of large stems. Major research efforts are needed in the study of stems in natural ecosystems, especially in the tropics and sub-tropics, and these must try to cover both the general surveys to find out what is there as well as more detailed studies to investigate microbial community structures and the interactions of populations within individual plants or plant communities.

Selected references and further reading

Ander, P. & Eriksson, K.E. (1978). Lignin degradation and utilisation by micro-organisms. *Progress in Industrial Microbiology*, **14**, 1–58.

Batra, L.R. (Ed.) (1979). *Insect-fungus symbioses; nutrition, mutualism and commensalism*. Allanheld, Osmun, Montclair. pp. 276.

Carroll, G.C., Pike, L.H. & Perkins, J.R. (1980). Biomass and distribution patterns of conifer twig microepiphytes in a Douglas fir forest. *Canadian Journal Botany*, **58**, 624–30.

Cotter, H. van T. & Blanchard, R.O. (1982). Fungal flora of bark of *Fagus grandifolia*. *Mycologia*, **74**, 836–43.

Dickinson, C.H. & Pugh, G.J.F. (Eds) (1974). *Biology of plant litter decomposition*, Vol. 1. Academic Press, London. pp. 241.

Eriksson, K.E. & Johnsrud, S.C. (1982). Mineralization of carbon. In Burns, R.G. & Slater, J.H. (Eds). *Experimental microbial ecology*. Blackwell Scientific Publications, Oxford, p. 134–53.

Frankland, J.C. (1982). Biomass and nutrient cycling by decomposer basidiomycetes. In Frankland, J.C., Hedger, J.N. & Swift, M.J. (Eds). *Decomposer basidiomycetes; their biology & ecology*. Cambridge University Press, Cambridge. p. 241–61.

Giles, K.L. & Atherly, A.T. (Eds) (1981). Biology of the Rhizobaiaceae. *International Review of Cytology, Supplement*, **13**. pp. 336.

Käärik, A.A. (1974). Decomposition of Wood. In Dickinson, C.H. & Pugh, G.J.S. (Eds). *Biology of plant litter decomposition*, Vol. 1. Academic Press, London. p. 129–74.

Liese, W. (Ed.) (1975). *Biological transformations of wood by microorganisms*. Springer-Verlag, Berlin. pp. 203.

Levy, J.F. (1982). The place of basidiomycetes in the decay of wood in contact with the ground. In Frankland *et al.* See above. p. 161–77.

Rayner, A.D.M. & Todd, N.K. (1979). Population and community structure and dynamics of fungi in decaying wood. *Advances in Botanical Research*, **7**, 333–420.

Roberts, W.P. (1982). Molecular basis for crown gall induction. *International Review of Cytology*, **8**, 63–92.

Round, F.E. (1981). *The ecology of algae*. Cambridge University Press, Cambridge. pp. 653.

Seaward, M.R.P. (Ed.) (1973). *Lichen ecology*. Academic Press, London. pp. 550.

Shigo, A.L. & Hillis, W.D. (1973). Heartwood, discoloured wood and microorganisms living in trees. *Annual Review of Phytopathology*, **11**, 197–222.

6
Microbiology of Roots

Root structure and chemistry

Roots have been studied for many years, from both morphological and physiological aspects, especially in relation to the growth and nutrition of economically important crop plants. The role of micro-organisms as a part of this system has been recognized from early this century when the rhizosphere was first defined as that volume of soil around legume roots in which growth of bacteria was stimulated. It is now applied generally to all plant groups and all micro-organisms around them. Microbes are an integral part of the root–soil system affecting both the root morphology and physiology. When considered critically the definition of the rhizosphere is not so clear cut: it has always been accepted that the outer limits are indefinite, depending on the plant species, soil type, water levels and many other factors. The inner limit was until recently defined as the root surface, the rhizoplane, near which the major part of microbial growth occurred. However, many studies have shown that epidermal and cortical cells die quite quickly and microbes, apparently saprotrophs, invade the cortex (p. 119). The interactions between microbes and roots are now considered to occur in the outer, invaded cortical layers, on the rhizoplane when this can be distinguished and also in the surrounding soil, the rhizosphere.

Roots form an unstable habitat for micro-organisms for the interfaces between the root, soil and microbes are continually changing. Roots grow by tip growth and are protected from abrasion by a mucilage cap and by expendable root cap cells: both these components are sloughed off and are major microbial resources. Behind the tip is a cell elongation zone, the root hair zone and then the mature epidermis and cortex. As noted above the cortex dies and in some roots secondary thickening occurs and bark and wood are formed. A root varies in age up its length and therefore has many different sorts of micro-habitats, both in terms of physical and chemical factors, available for a wide range of microbes. Alternatively, the process can be considered in terms of a single position on the root which ages and changes with time. Healthy, vigorous root tips elongate so rapidly that bacteria and fungi cannot grow fast enough to colonize them from pre-existing root parts and they are colonized from the soil. Only protozoa may be able to move fast enough to keep pace with the growing tip.

There have been many reviews of the microbiology of the rhizosphere, a whole book is devoted to the soil–root interface and many chapters in books on soil biology. Some of the more recent or comprehensive ones are: Foster,

Rovira & Cock (1983) on the structure, ultrastructure and chemistry of the root surface; Harley & Scott Russell (1979) on all aspects of the soil–root interface; and Lynch (1983) on the microbiology of the root surface.

Root structure obviously varies greatly between species, not only in size and extent but also in basic growth patterns. Some plants form roots from the seed, seminal roots, which grow and branch to form the entire system. Other plants, such as grasses and cereals, produce seminal roots and also adventitious roots at the stem nodes, nodal roots, so that they can produce a new root system if the existing one is damaged. There is a great turnover of root material in natural soil conditions with old roots dying and new growing points being produced.

The root tip is covered by mucilage, a gel of mixed acidic polysaccharides, within which is embedded root cap cells which have been sloughed off. These cells may remain alive for some time (Fig. 6.1A) but as the root tip moves on and the tissues beneath the cells differentiate, they die. This may however take several days and free, live cells, probably from the root cap, are found amongst the root hairs. A healthy maize plant will slough up to 10 000 cells a day which represents a considerable expenditure of resources by the plant, and a useful source of nutrient to micro-organisms.

Root surface cells, usually epidermis in young roots, have a layer of mucilage closely attached to their surface. As the wall area rapidly increases in the elongation zone this layer may become very thin except over the cell junctions (Fig. 6.1B). In addition to this firmly attached inner mucilage there is also a more tenuous layer, which extends into the soil, binding particles together and also containing micro-organisms. Some of this mucilage may be mixed with capsular and other slimes of microbial origin and it usually appears fibrous after fixation and dehydration for microscopy. The origin of this more diffuse mucilage layer is in doubt and may vary from species to species. In maize it may arise from the root cap or from the cells derived from the root cap. Alternatively it may be the result of the breakdown of outer mucilaginous layers of the epidermal cell walls.

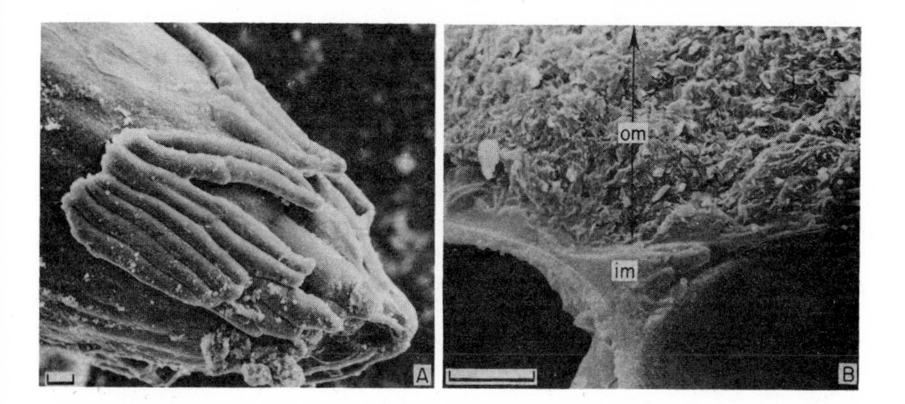

Fig. 6.1 **A.** Live root cap cells in mucilage. Scanning electron microscope, frozen-hydrated preparation. Bar = 10 μm. **B.** Scanning electron micrograph of frozen-hydrated material to show inner (im) and outer mucilage (om) layers. Bar = 5 μm.

The mucilage layers around roots are important to the plant because they help bind soil round the root and maintain good continuity with the water films over the roots and round soil particles. Root hairs are also important in binding soil and in maintaining close contact between the root and the soil. In grass roots there seems to be good evidence of a discrete soil sheath around the roots held by root hairs and mucilage.

The water potential in the soil around roots is usually not much reduced, compared with un-rooted soil, provided that there is continuity between the soil water and the plant root, and the water enters uniformly over the root surface. However these last two assumptions may not be valid. It is known that different plants form roots at different depths and therefore do not draw water from all positions down the profile. Root-soil contact may not be good because channels form in the soil round roots, or roots grow through voids. Roots also shrink under water stress. There may therefore be local water stress in the rhizosphere if water uptake is concentrated into a rather small proportion of the total root which is in direct soil contact. There may also be some compaction of soil around roots, generated when the root pushed through the soil. This in itself restricts water flow. If the potentials are no lower than in the root xylem, i.e. supposing a continuous contact, then it may go to 2000 KPa (= −20 bars) and lower in the rhizosphere of xerophytes. This is below the water potential at which many fungi and bacteria will grow, but not less than occurs in normal soil under dry conditions.

Gas exchange in the rhizosphere is complex. In well-aerated soils there may be slight elevation of carbon dioxide levels and depletion of oxygen. However as soon as the soil becomes waterlogged or even wet there is a different situation because of the very slow diffusion of oxygen and carbon dioxide through water films around roots and soil voids. Oxygen is decreased, carbon dioxide is elevated, methane and ethylene occur and may reach physiologically-important levels. These changes may be caused by the root alone, by soil organisms alone or by a combination of the two. The situation becomes complicated in persistently poorly-aerated soils because the roots of many plants form aerenchyma, open channels down the root cortex, which allows gaseous diffusion. The root then becomes a source of oxygen, not a cause of anoxia. The formation of aerenchyma may be caused by the presence of ethylene. The level of aeration also affects other microbial activities, especially nitrogen conversions (p. 123).

The major change in the rhizosphere, compared with ordinary soil, is the increase in organic matter. The organic matter takes many forms, but some care is necessary in the interpretation of the data in the literature. Most detailed chemistry is based on results from young, often axenic plants. There is very little information from natural, annual plants, let alone trees, growing in anything approaching normal field conditons. Several categories of compounds are however recognized: (1) Exudates, usually low molecular weight organic compounds and minerals which passively leak out of the cells. (2) Secretions which are actively released by the roots, e.g. some enzymes. (3) Mucilages from the root cap or epidermal cells as discussed above, which may be difficult to assess and analyse because under natural situations they become mixed with microbial capsular materials. (4) Cells or their parts (walls or cytoplasm) which are shed from the root. This may be root cap cells or their contents and eventually the walls of dead cortical cells. The amounts of materials involved are very difficult to measure for there is a basic methodological problem. If the plants are not axenic then the micro-organisms will change and respire the exudates, etc. On the other hand in the absence of micro-organisms there is probably less production of materials and the situation is unnatural. It is possible that microbes act as the sink for soluble exudates, maintaining a concentration gradient and so promoting further release. Microbes could also cause or hasten cortical cell death and may affect hormone balance (p. 119) and hence cell permeability. However, by the use of radio-active tracers, mostly ^{14}C, and various approximations to natural conditions some figures have been produced. In wheat and barley grown in sterilized soil about 60 to 77% of the carbon translocated to the roots is incorporated into the root system, 3–12% is respired, 17–25% appears as insoluble material in the soil and 3–9% in water-soluble exudates. Some of the insoluble material (probably quite a lot) could be dead or dying roots, as well as sloughed cells and mucilage. Other experiments show values for soluble exudates of 1–10% of the root weight and insoluble material, mostly mucilage in the primary roots, as less than 1% of the root weight. These data are all for actively growing plants. As the root ages there is thought to be less soluble exudates and more insoluble material, especially sloughed cells, but there is no clear evidence from older roots, or roots of any age growing in normal soil.

The chemical nature of this organic matter has been determined, though again usually for sterile plants. The sloughed cells contain normal cell constituents and walls are made mostly of cellulose. The chemistry of the mucilage has been discussed above. The soluble materials are very diverse but are mainly low molecular weight amino acids (methionine, proline, glutamic acid, alanine, asparagine) and simple sugars (glucose, mannose, galactose). Organic acids, vitamins, nucleotides, enzymes, foliar applied pesticides, etc., have all been reported, and there are also volatiles such as terpenes, ethanol, isobutanol and acetoin. Under more natural conditions with mature plants measurements are made on a different basis: the sum total of plant and microbial products is measured and compared with that in non-rhizosphere soil. Microbial cells would also be included. Rhizosphere soil contains more soluble sugars but less insoluble material than the surrounding soil and there is less nitrogen but more polyphenols. There are also variations with plant age, with more sugars from older plants. Such data are difficult to interpret because of the complexity of the system. They certainly do not represent what is exuded by the plant but they do show that, no matter whether soluble carbon comes from the plant direct or from microbial breakdown of insoluble materials, there are significantly more microbial substrates in the region of the roots. This is of great importance for in many agricultural situations soil is substrate limited for microbial growth. One of the main effects of roots is therefore to increase the amount of microbial growth.

Microbial saprotrophs, their numbers and activity near roots

There are now adequate data on microbial numbers, biomass and activity in soil (Campbell, 1983) and near roots there is a general increase in all these parameters. The increase in microbial numbers near the root is often expressed as an R:S ratio, the ratio of numbers in the rhizosphere to numbers in root-free soil. This is frequently as high as 10, and may be hundreds or even thousands for particular groups or taxa of bacteria (Table 6.1). Notice that different plant species have different bacterial numbers (and fungal numbers) and different rhizosphere effects. These characteristics can also be seen in Tables 6.4 and 6.6. Not all micro-organisms respond similarly, and indeed some particular

Table 6.1 Colony counts of bacteria in root-free soil and in the rhizosphere of crop plants and wild species. The last three species are sand-dune plants and so have low numbers but a very pronounced rhizosphere effect. (From Woldendorp, 1978.)

Species	**Colony counts of bacteria** (10^6 g^{-1} soil)		
	Rhizosphere	Root-free soil	R : S ratio
Trifolium pratense	3260	134	24
Avena sativa	1090	184	6
Linum usitatissimum	1015	184	5
Zea mays	614	184	3
Atriplex babingtonii	23.3	0.016	1455
Ammophilia arenaria	3.58	0.016	223
Agropyron junceum	3.56	0.016	222

organisms may decrease near the root (Campbell, 1983). In very densely-rooted systems, such as intensive agricultural crops or grassland (Table 6.3) all the soil may effectively be rhizosphere, there being no point not under the influence of roots. Where plants are more widely spaced, and especially where environmental conditions impose stress on vegetation such as in dry or cold deserts or in arid regions, the plant density will be too low to affect most of the soil.

Amongst bacteria it is widely accepted that Gram-negative rods, especially *Pseudomonas*, show the greatest increases amongst the general saprotrophs and this certainly seems to be true in young plants (Tables 6.2 and 6.3). However in older plants (Table 6.2), or stable plant communities (Table 6.3) the coryneforms become more important. The effect of plant age on bacterial numbers is further illustrated in Fig. 6.2. It may not be the chronological age so much as the physiological age that is important; the peak numbers appear near to flowering time in those few plants that have been investigated. It is also possible that the now standard culture techniques may be imposing a distorted picture: many unlikely organisms can be shown to occur in the rhizosphere in considerable numbers if correct isolation procedures are used. For example the photosynthetic bacteria, such as the *Rhodopseudomonas*, can be isolated and are apparently living heterotrophically in the rhizosphere.

Table 6.2 The influence of plant age on the proportions of Gram-negative rods and coryneforms in the rhizosphere of flax. (From Woldendorp, 1978.)

Plant age (weeks)	**G (–) rods** (%)	**Coryneforms** (%)	**Other micro-organisms** (%)
1	90	4	6
4	81	14	5
8	60	32	8
12	49	48	3

Table 6.3 Composition of the microflora of the rhizosphere of a 2 year-old sand culture of perennial rye-grass, of a permanent grassland on sand and of a sandy arable soil. (From Woldendorp, 1978.)

	Rhizosphere	**Grassland**	**Arable soil**
% of total number			
Bacteria	75	70	90
Streptomycetes	25	30	9
Fungi	0	0	1
% of bacteria			
Gram-negative rods	60	58	11
Bacilli	3	10	14
Coryneforms	37	32	75

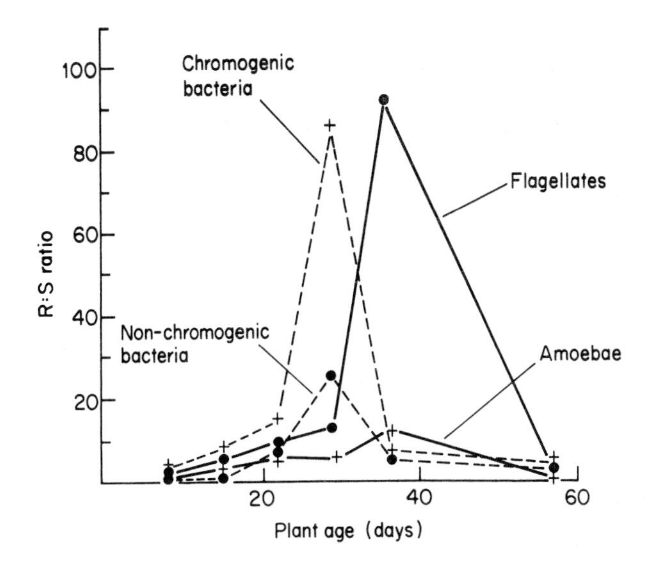

Fig. 6.2 The rise in bacterial and protozoan numbers in the rhizosphere of *Sinapis alba* (L.) (white mustard). The R:S ratio is plotted against plant age: the maximum rhizosphere effect for protozoa is later than for bacteria. (From Campbell, 1983; based on data from Derbyshire J.F. & Greaves, M.R., 1967, *Canadian Journal of Microbiology*, **13**, 1057–68.)

Table 6.4 Abundance of bacteria and fungi on the root surfaces of *Lolium perenne* and *Plantago lanceolata*, determined by direct observation and plate counts. (From Rovira, A.D., Newman, E.I., Bowen, H.J. and Campbell, R., 1974, *Soil Biology and Biochemistry*, **6**, 211–16.)

	Lolium	Plantago	P*	LSD
Direct observation				
Fungal hyphae, length, mm mm^{-2}	12.1	14.3	–	3.6
Bacteria, frequency %	66.7	66.2	–	6.6
Bacteria, cover %	7.7	6.3	<0.05	1.3
Bacteria, number mm^{-3} root × 10^3	1090	990		
Plate counts				
(number mg^{-1} root fresh weight × 10^3)				
'Total' bacteria	107	70	–	50
Gram-negative bacteria	48	26	<0.05	21
Filamentous fungi	0.35	0.32	–	0.21
Yeasts	2.6	0.95	<0.1	2.0
(number mg^{-1} root dry weight × 10^3)				
Total bacteria	737	621	–	368
Gram-negative bacteria	330	214	–	157
Filamentous fungi	2.3	2.8	–	1.6
Yeasts	18.1	8.2	–	14.2

* Statistical significance of difference between the species, by unpaired *t* test; $P > 0.1$ is shown as –.
LSD value at $P = 0.05$.

Rhodopseudomonas will not be isolated however on media designed for true heterotrophic bacteria.

Even though the reported numbers of bacteria in the rhizosphere seem very large they do not cover much of the root surface, usually less than 10% (Table 6.4 and 6.5)

Table 6.5 Range of microbial abundance in pots in a glasshouse, expressed in various units. (From Christie, P., Newman, E.I. & Campbell, R., 1978, *Soil Biology and Biochemistry*, **10**, 521–7.)

	Bacteria		Fungi	
	maximum	minimum	maximum	minimum
Length per unit root-surface area (mm mm^{-2})			15.5	0.73
Cover (%)	7.7	0.24	3.9	0.18
Microbial volume per unit root-surface area (μm^3 μm^{-2})	4×10^{-2}	0.1×10^{-2}	8×10^{-2}	0.4×10^{-2}
Microbial dry weight per unit root-surface area (ng mm^{-2})	8	0.3	15	0.7
Microbial volume per unit root volume (μm^3 μm^{-3})	8×10^{-4}	0.2×10^{-4}	14×10^{-4}	0.7×10^{-4}

The fungi are numerically much less important than the bacteria, even several orders of magnitude less so that as a percentage of the total number of microbes they become very small indeed (Table 6.3 and 6.4). This is a reflection of the difficulty of getting meaningful data from dilution plates for numbers of fungi (p. 11). If their size and frequency are determined by direct observation they have as high a biomass and occupy as much or more of the surface as do bacteria (Table 6.5).

Protozoa occur in the rhizosphere (Fig. 6.2) in response to the elevated numbers of bacteria and they show R:S ratios of up to 10. The species present are the same as in ordinary soil and they mostly are bacterivorus (e.g. *Bodo, Cercommonas, Colpoda*, small amoebae). Their feeding is selective, under laboratory conditions anyway, and they may therefore affect not only bacterial numbers but also species composition. Protozoa increase mineralization rates near roots. Myxomycetes also occur and may have R:S ratios of 10–100 (Table 6.6). Little is known of their importance but the myxamoebae may be up to one third of the total amoebae in some cases, though often much less, and they are active predators on bacteria. Again there are variations in numbers between different host plants.

Trying to measure the activity of microbes near roots is difficult because there are technical problems in separating microbial and root respiration, but this is possible using ^{14}C with suitable controls. Most studies show that 30 to 50% of total respiration is due to micro-organisms and the rest to the roots, assuming sterile roots respire at the same rate as natural ones. This ignores metabolism by anaerobic microbes. However even with these provisos there is obviously considerable activity in the rhizosphere and this is in contrast to the situation in normal soil.

Individual plants can be studied by pulse labelling or continuous application of radio-active carbon sources, usually supplying $^{14}CO_2$ to the photosynthesizing tops and then looking at the root exudates or the surrounding microbes or in the CO_2 from root respiration for the radio-activity. Such data are presented in Fig. 6.3; there is an extra peak in the specific activity and the $^{14}CO_2$ evolution of the non-axenic plant which is interpreted as being the microbial respiration of the exudate.

The activity of particular groups of bacteria has also been studied, especially those concerned with the nitrogen cycle. The nitrogen cycle has many interrelated reactions (Campbell, 1983) whose effect on plant nutrition is considered later (p. 122) but we may note that nitrogen fixation activity may be increased, there may be a decrease in the activity of nitrifying bacteria and an increase in the immobilization of ammonia.

There is also an increase in the activity of some enzyme systems around the root (e.g. phosphatase) and, while some of this is certainly of plant origin, it could also be produced by microbes on the rhizoplane.

The distribution of saprotrophic microbes on the root surface

The micro-organisms are not uniformly distributed over the root of any plant. They may occur within the mucilage layers (Fig. 6.4A), usually the outer layer or on the mucilage surface, and the hyphae of fungi pass out into the soil (Fig. 6.4B). The microbes may be attached to the root surface by polysaccharide or

Table 6.6 Numbers g^{-1} fresh weight, of myxomycetes, myxobacteria, dictyostelids and protozoa in the rhizosphere of sand dune plants. R:S ratios are given in parenthesis below each number. Some of the very high R:S ratios for *Raphanus raphanoides* are caused by some roots in the sample having necrotic lesions. (Data from Alan Feest, Department of Botany, University of Bristol.)

Micro-habitat	**Amoebae including myxomycetes**	**Myxomycete amoebae**	**Myxomycete plasmodium-forming-units**	**Dictyostelids**	**Myxobacteria**	**Ciliates**
Sand	1775	114	83	0	261	208
Rhizosphere of *Carex arenaria*	17658 (9.9)	6558 (57.5)	1261 (15.2)	0	4036 (15.5)	12612 (60.6)
Rhizosphere of *Tripleurospermum maritimum*	11657 (6.6)	466 (4.1)	366 (4.4)	0	3664 (14.0)	2331 (11.2)
Rhizosphere of *Raphanus raphanoides*	155769 (87.8)	1038 (9.1)	892 (10.7)	35	276896 (10 61)	9371 (45)

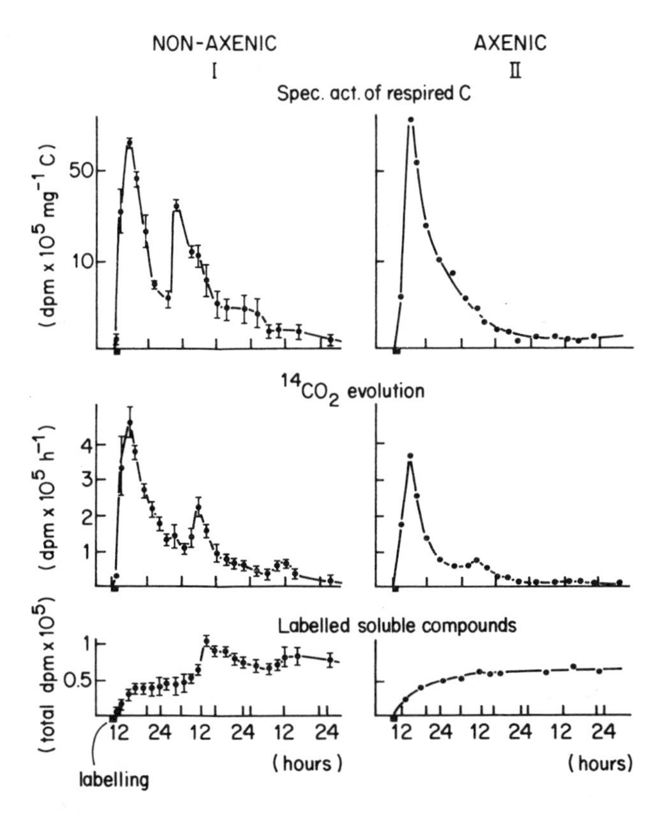

Fig. 6.3 Simultaneous variations of specific activity of CO_2, $^{14}CO_2$ fluxes and labelled soluble substances liberated in the axenic and non-axenic rhizosphere of wheat labelled with $^{14}CO_2$. (Plants grown in nutrient solution with 9 h dark at 18°C and 15 h light at 28 °C with a light intensity of 30 000 lux.) (From Waremburg, F.R. & Billes, G. in Harley and Scott Russel, 1979.)

glycoprotein, at least they appear so if material is fixed for electron microscopy: more usually however they are mixed with the adhering soil crumbs and clay particles. Clay platelets are usually adsorbed on to the microbial surface edge-on. The positive edge of the clay being held by electrostatic attraction to the predominantly negative cell wall.

Considering the root as a whole, there are very few organisms on the root cap (Fig. 6.1A), except occasional ones that happen to be touched as the root moves through the soil, and there is little evidence of bacteria and fungi until after the zone of elongation (Fig. 6.5). Around the root hair zone there are many bacteria, often in small colonies of less than 100 cells with an uneven distribution in relation to the epidermal cells. Many studies, but not all, have shown that bacteria tend to grow most over the cell junctions, in the depressions on the root surface between the bulging epidermal cells. In contrast to the leaf surface (p. 53) hyphae of fungi do not show such a strong tendency to follow the depressions on the root.

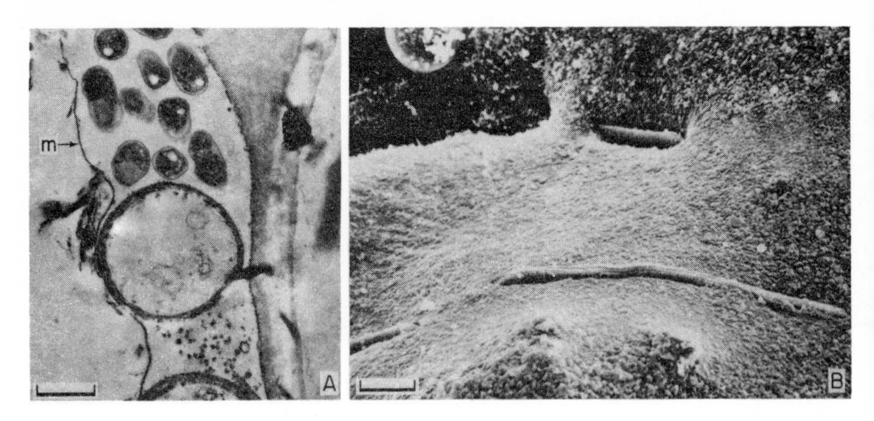

Fig. 6.4 **A.** Bacteria and a lysed hypha under the mucilage layer (m) around the root. Bar = 1 μm. (From Faull, J.L. & Campbell, R., 1979, *Canadian Journal of Botany,* **57**, 1800–8.) **B.** Scanning electron micrograph of the surface of the mucilage around a root with the hypha dipping in and out of the mucilage layer. Bar = 10 μm. (From Campbell, R., 1983, *Canadian Journal of Microbiology,* **29**, 39–45.)

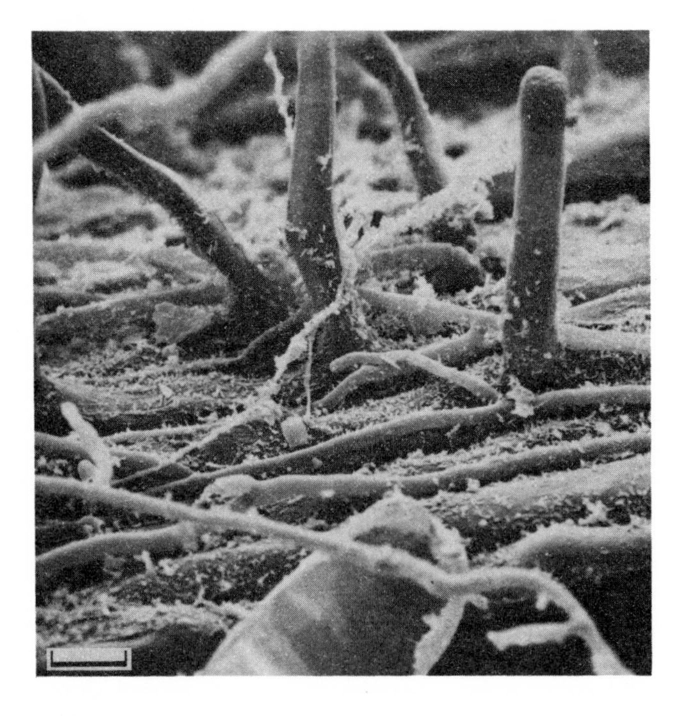

Fig. 6.5 The root hair zone with an unusually dense colonization by bacteria and hyphae. See also Fig. 6.18 B. Bar = 10 μm.

The rate at which microbes grow on root surfaces is difficult to assess under anything like natural conditions but estimates can be made from populations on roots of known ages. Doubling times for bacteria are only a few hours under optimum conditions, though there is great variation between bacteria. From experimental studies it seems that a given root may be able to sustain a given biomass and the bacterial population on the root surface may stabilize at this, even if different inoculum sizes are involved (Fig. 6.6). This is also predicted for the rhizoplane by the computer model produced by Newman & Watson (1977, see Newman, 1978), but away from the root the initial inoculum density has a great effect on the final population in this simulation.

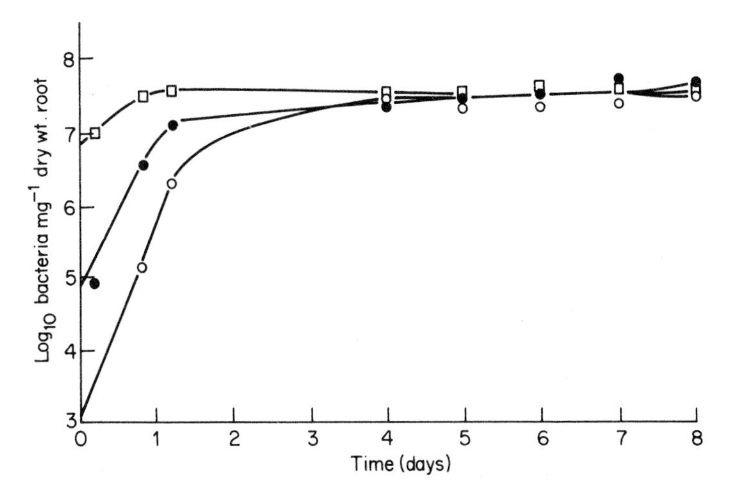

Fig. 6.6 Effect of original inoculum size on the growth rate and final population size of a fluorescent *Pseudomonas* sp. on barley roots. (Based on Bennett, R.A. & Lynch, J.M., 1981, *Current Microbiology,* **6**, 137–8.)

The distribution of micro-organisms therefore varies over a short distance between different cells on the root and in general up the root as it ages. Does this represent different colonization patterns, different growth patterns or preferential movement of organisms or a combination of all these factors? If roots are artificially inoculated with an approximately uniform cover of organisms they grow most over the cell junctions, suggesting that these aggregates are due to preferential cell growth rather than preferential inoculation. Possibly the cell junctions are favourable positions for the exudation of soluble material.

Experimental studies with inoculation of bacteria singly or in combination onto sterile roots show that the growth rate and ability to colonize is subject to antagonism by other microbes (see also p. 150). Some bacteria that are able to colonize the root on their own are suppressed or even eliminated when inoculated in a mixture. Similarly bacteria or fungi may inhibit fungal growth on the roots.

The overall effect is that there is great variation between individual roots in the numbers of bacteria at different distances from the tip. This presumably

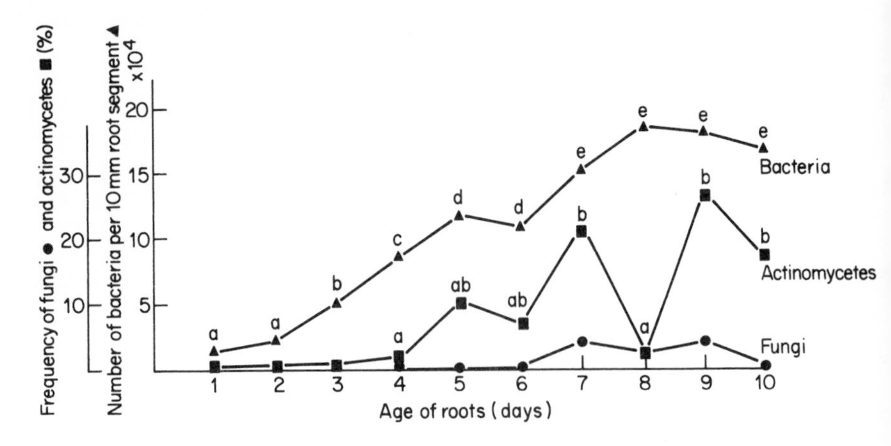

Fig. 6.7 Colonization of seminal roots of wheat by bacteria (excluding actinomycetes), by actinomycetes and by fungi. Subsequent points on the same curve with different letters are different at P = 0.05. (Based on graph and data of van Vuurde, J.W.L. & Schippers, B., 1980, *Soil Biology and Biochemistry,* **12**, 559–65.)

reflects chance encounters with different soil populations or is possibly the result of random damage to the roots giving locally high numbers. Figure 6.7 shows that bacterial numbers increase from the root tip, then they stabilize but increase again to a maximum as the root ages. Fungi only appear in any number after bacterial colonization. In this example the initiation of lateral roots started at the time of greatest increase but was unaltered after day 5. Measurements of epidermal and cortical cell senescence showed a similar trend, increasing from the root tip, until day 5, remaining steady until day 7 and then dropping to day 10 when approximately two thirds of the cells were senescent. Cortical cell death and the production of lateral roots would lead to increases in available nutrients.

The rate at which senescence occurs varies with the plant and the environment. Some authors suggest that the cortical and epidermal cells die in a matter of a few days, as above, in other plants the roots seem to remain intact for weeks or even months. Superimposed on this senescence of particular cells or a particular root is a variation in the general pattern of plant root activity or growth over a season and with the depth in the soil, which probably reflects not only the growth stages of the plant but also the soil moisture and temperature. There are no data available for the effect of these changes on micro-organisms on the roots of known ages, but general seasonal variation in activity has already been noted (Fig. 6.2). There is also a continual turnover of roots in the soil, with many dead roots (up to half the root biomass) even in a healthy, actively-growing plant community. There are flushes of fine root growth during favourable periods and then many of these roots die, are decayed and the nutrients are released. Roots may go back down the same root run, or certainly through the same soil volume at the next growth period to re-use the nutrient. There have been some studies on the colonization and decay of these roots (Waid, 1974). The main colonizers of the dead cells are fungi, usually ones which can degrade cellulose. There is progressive colonization of the inner cortex and the stele, non-spore forming strains of fungi (sterile mycelium) penetrating deepest. The initial colonizers of the outer cortex (*Penicillium*, *Gliocladium*, *Fusarium* and zygomycetes) do not penetrate into the inner cortex and stele (Fig. 6.8). A similar study on *Phaseolus* roots also

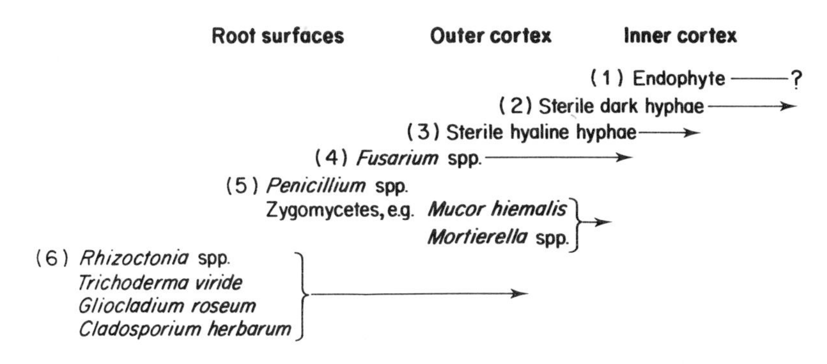

Fig. 6.8 Colonization of root surfaces and the cortex of *Lolium perenne* by different groups of fungi. (From Waid, J.S., 1957, *Transactions of the British Mycological Society*, **40**, 391–406.)

showed that penetration of some surface fungi was limited (Table 6.7) and that the stele was colonized predominantly by *Cylindrocarpon* and sterile dark hyphae. A rather different technique is to artificially kill the roots by cutting off the top of the plant and then to study the colonization of the healthy roots as they decay (Table 6.8). Clearly some fungi are restricted to the healthy root surface (e.g. the first four isolates) and others such as some *Fusarium* sp. occur at all stages of decay. As shown in Table 6.8 there are isolates only found in the initial decay stages (from *Pythium debaryanum* on the list down to *Trichoderma glaucum*) and also those only found late in the decay (*Sporotrichum*, *Papularia*) or those that increase towards the end (*Humicola grisea*, *Gliocladium roseum*).

Information on bacteria from dead and decaying roots is almost non-existent. They do occur in all parts of the roots but information on numbers or frequency and the types present is not available. It is probable that they are as important in the decomposition as they are for leaves or wood and micro-arthropods and protozoa presumably also play a role.

Effects of saprotrophs on the plant

It is quite possible to grow plants in gnotobiotic culture; growing with known combinations of organisms with or without soil. Indeed without micro-organisms plants usually have longer roots and more root hairs than in natural conditions. Addition of selected microbes can reduce root length (Table 6.9). Many microbes isolated from the rhizosphere have been shown to produce substances in culture with activity analagous to plant hormones (auxins,

Table 6.7 Fungi associated with dwarf bean roots showing the estimated relative importance of each species in each of four easily recognizable habitats. (From Taylor, G.S. and Parkinson, D., 1965, *Plant and Soil*, **22**, 1–20.)

Fungus	Root surface	Cortex	Outer stele	Inner stele
Mucor spp.	*	*		
Mortierella vinacea	*	*		
Trichoderma viride	**	*		
Fusarium sambucinum	**	*		
Penicillium spp.	**	**		
Penicillium lilacinum	**	**		
Mortierella spp.	**	**	*	
Gliocladium spp.	***	**	*	*
Fusarium oxysporum	*****	*****	*	*
Cylindrocarpon radicicola	***	***	*****	****
Sterile dark forms	*	*	****	*****
Varicosporium elodea		*	**	**
Sphaeropsidales		*	*	
Botrytis cinerea		*	*	
Sterile hyaline forms			*	**

* usually present but only low frequency of isolation.
***** dominant form with high frequency of isolation.

gibberellins and cytokinins). Production has not been shown to occur in natural soil, but saprotrophic microbes may well affect root morphology and growth and also the growth and yield of the whole plant. Such effects are now being exploited in the development of plant growth promoting organisms for commercial inoculation. Significant increases in yield can be obtained with inoculation with selected bacteria. One of the problems however is to get them into the soil or onto the roots in sufficient numbers. Even then it may be difficult to produce consistent results both in trials in the same year and with the variation from year to year. Just how these bacteria increase growth is not clear. Some may control minor plant diseases or may produce the hormones discussed above. Another possibility is the production of iron-chelating chemicals (siderophores) which reduce the level of Fe^{3+} to such an extent that other microbial growth is inhibited. This includes inhibition of known pathogens and also of bacteria and fungi which reduce growth (Table 6.9).

There is remarkably little evidence, though much speculation, on the effects of saprotrophs on the availability of mineral nutrients. There are isolated reports of increases or decreases in the uptake of this or that mineral, but no coherent picture. For example, uptake of managanese may be reduced by microbial oxidation, but microbial-chelating agents may assist in making manganese soluble. There could also be pH effects on solubility, but the microbial contribution to such changes is likely to be small. Phosphorus has been especially studied, probably because it has an easily used isotope rather than because of its intrinsic importance. Microbes produce phosphatases, as does the root, which may solubilize phosphorus in the soil and there is some evidence from artificial systems that in young plants microbes could increase ^{32}P uptake and ^{32}P translocation to shoots (Table 6.10). There can however also

Table 6.8 Fungi occurring on tap roots of *Vicia faba* before and after excision of the root system. (From Waid, 1974; Based on data of Matiques, P.L.J., 1966, *School Science Review*, **48**, 108–23.)

Fungal frequency per 100 root segments	Before excision	Days after excision					
		3	5	7	10	14	21
Aspergillus fumigatus	18	0	0	0	0	0	0
Aspergillus sulphureus	15	0	0	0	0	0	0
Mucor corticolus	24	0	0	0	0	0	0
Gliocladium fimbriatum	15	0	0	0	0	0	0
Pythium debaryanum	52	48	42	48	33	30	30
Mucor globosus	45	45	39	30	21	5	9
Penicillium janthinellum	55	42	12	3	0	0	0
Trichoderma viride	30	18	15	6	0	0	0
Pencillium spp. (Asymmetrica)	8	15	0	3	0	0	0
Trichoderma koningii	27	15	12	12	0	0	0
T. album	8	15	9	0	3	0	0
Aspergillus violaceo-fuscus	15	15	5	0	0	0	0
Cephalosporium curtipes	8	12	15	9	9	0	0
Cylindrocarpon spp.	24	12	0	3	0	0	0
Mucor jansseni	10	10	0	0	0	0	0
M. varians	12	10	0	0	0	0	0
Trichoderma glaucum	10	9	6	0	0	0	0
Penicillium spp. (Monoverticillata)	0	6	0	0	0	0	0
Mucor hiemalis	0	6	0	0	0	0	0
Verticillium terrestre	5	6	6	0	3	0	0
Mortierella sp.	0	6	0	3	0	6	0
Rhizoctonia solani	6	6	6	6	0	3	0
Penicillium nigricans	0	3	0	0	0	0	0
Mucor albo-ater	0	0	0	0	6	15	15
Botrytis cinerea	15	7	0	12	18	12	6
Gliocladium catenulatum	15	0	0	6	6	0	0
Volutella ciliata	6	0	3	12	27	24	0
Aspergillus koningii	0	0	0	0	6	9	0
Sporotrichum epigeum	0	0	0	6	12	24	6
Papularia sphaerosperma	0	0	0	6	9	9	0
Fusarium oxysporum	75	75	72	80	84	88	92
Fusarium solani	46	38	36	44	48	55	72
Humicola grisea	20	24	28	36	47	59	82
Gliocladium roseum	28	24	28	36	55	66	69
G. penicilloides	0	0	0	0	9	15	18
Dicoccum asperum	6	0	0	0	0	6	24
Trichothecium roseum	0	0	0	0	0	0	15
Chaetomium globosum	8	0	6	9	6	15	24
Papulospora sp.	15	0	0	9	9	15	27
Spicaria sp.	8	0	0	6	0	12	9
Stysanus stemonites	0	0	0	0	3	9	18
Stemphylium sp.	14	0	0	0	3	18	9
Average number of fungi per root segment	6.4	4.7	3.4	3.8	4.2	5.0	5.3

Table 6.9 Microbial competition and barley growth. Results are expressed as a percentage increase or decrease over the control to which no micro-organisms were added. (From Lynch, J.M. & White, N., 1977, *Plant and Soil,* **47**, 161–70.)

Micro-organism	**Inoculum concen-trations** ($g\ l^{-1}$)	**Length of longest leaf** (%)	**Number of seminal roots** (%)	**Total length of seminal root axes** (%)	**Length of longest root** (%)
Azotobacter chroococcum	0.5	+0.4	0	−24.7*	− 5.1
Blakeslea trispora (+)	1.3	−2.6	+6.6	−34.2*	−29.7*
A. chroococcum + *B. trispora* (+)	0.5 − 1.3	+6.3	−6.6	−24.4*	−16.0

* Results significantly different from control at $P = 0.05$.

be competition for phosphorus, the microbes immobilizing it and making it unavailable to the plant. Notice however that these plants are still only 12 days old: in soil experiments with plants up to ten weeks old there was still less phosphorus absorbed by the non-sterile plant, suggesting that if phosphorus is limiting there could be competition between the roots and the microbes. If there is competition the saprotrophic microbes win! The situation with biotrophs is different, and is considered later (p. 125).

Availability of nitrogen is also affected by biotrophs, especially *Rhizobium* (p. 113) but there is also a considerable effect on the uptake of nitrogen which is attributable to saprotrophs. Micro-organisms in the rhizosphere can produce ammonium by degradation of organic material. This mineralization makes inorganic nitrogen available to the plant. It has been shown that protozoa grazing on the bacteria can significantly increase the rate of mineralization and uptake. Nitrification ($NH_4^+ \rightarrow NO_3^-$) in the rhizosphere has been much

Table 6.10 Effect of soil micro-organisms on the uptake of phosphate by barley seedlings of different ages. The plants were grown in aerated complete nutrient solution and uptake measured after treatment for 30 minutes in 0.005 mM potassium phosphate labelled with ^{32}P (From Agricultural Research Council, Annual Report *Letcombe Research Station,* Wantage, England, 1974.)

Age	**Day 6**		**Day 8**		**Day 12**	
Condition of plants	**Sterile**	**Non-sterile**	**Sterile**	**Non-sterile**	**Sterile**	**Non-sterile**
Roots						
Dry weight (mg)	6.0	5.5	7.5	6.7	10.4	8.8
Phosphate absorbed ± S.E. ($pmole\ mg^{-1}$ dry wt)	559±28	1045±53	631±36	1004±61	584±24	908±48
Shoots						
Dry weight (mg)	13.2	15.1	17.5	19.6	27.0	26.9
Phosphate absorbed ± S.E. ($pmole\ mg^{-1}$ dry wt)	19±1	40±8	21±1	31±2	15±1	12±1

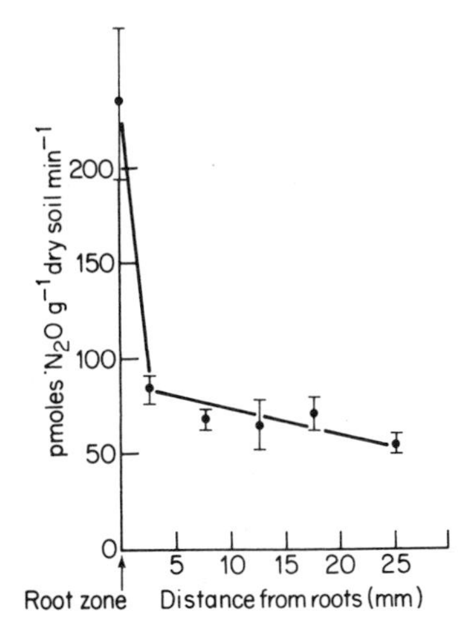

Fig. 6.9 Denitrifying activity of soil sampled from the root zone of oats at various distances from the root. (From Smith, M.S. & Tiedje, J.M., 1979, Reproduced from *Soil Science Society America, Journal,* **43**, 951–5 by permission of the Soil Science Society of America.)

disputed. It may not occur in grassland or woodland systems simply because nitrogen is limiting and free ammonium is not available as a substrate. Ammonium is taken up by plants and/or microbes as soon as it is released by mineralization. There is also the possibility that plants release nitrification inhibitors, but this is not proven. Where excess nitrogen is applied as fertilizer in agricultural systems the situation may be different. It may not however be desirable to have nitrification since fertilizer applied as ammonium is adsorbed on clay particles, whereas nitrate fertilizer or nitrate produced by nitrification is subject to rapid leaching. Nitrification-inhibiting chemicals are commercially available to prevent the loss of nitrogen after the application of ammonium fertilizers or in situations where nitrification is likely to be significant.

Denitrification ($NO_3 \rightarrow N_2$ or N_2O) may be a serious problem because it results in a loss of nitrogen from the system. It occurs especially under conditions of poor aeration when nitrate is used as a terminal electron acceptor rather than oxygen. Roots and microbes reduce oxygen levels and it has been shown by many authors that denitrification is increased in the rhizosphere of many plants. For example in grassland, the soil of which is almost all rhizosphere since the root density is so high, the amount of added nitrate–nitrogen lost was 15–37%, being greatest in clay and wet peat soils with the worst aeration (Table 6.11). Denitrification does of course occur in the soil in general, as well as associated with plant roots, but it is much more active near roots (Fig. 6.9) and is detrimental because of the loss of available nitrogen. If the soil is very poorly aerated and the redox potential drops below −220 mV then sulphate reduction can occur and the hydrogen sulphide produced is highly toxic and may cause damage to roots (of rice for example).

Nitrogen fixation ($N_2 \rightarrow NH_3$) is carried out only by prokaryotes which may be symbiotic (p. 125) or free-living. Nitrogen fixation requires a considerable amount of energy (Postgate, 1982) and it has been calculated that in bulk soil there is not enough available carbon. However round the roots there may be sufficient exudate to allow some fixation of nitrogen, the products of which could be available to the plant. Attention has recently been focussed on the colonization of some tropical grass roots by *Azotobacter paspali, Azospirillum brasilense* and *Beijerinckia* sp. These bacteria are found especially in the rhizosphere and colonize the mucilage layer and they penetrate between the root cells. They are all capable of fixing nitrogen, the question is do they fix nitrogen in the soil, and if so is that fixed nitrogen available to the plant? The first answer is a qualified yes; the doubt is whether the quantities fixed are significant in relation to that available from mineralization. There is no proof that any nitrogen that is fixed actually gets to the plant. However detailed nitrogen balances often do show a small net gain. This amount may be significant over long time periods in natural ecosystems to balance losses by leaching and denitrification, or it might make all the difference in very nitrogen-deficient habitats like sand dunes. However in western 'developed' agricultural situations with the continual removal of high yields this non-symbiotic nitrogen fixation is probably insignificant in the amounts shown by recent estimates.

Table 6.11 Distribution of fertilizer nitrogen after addition of labelled nitrate to grassland sods. (From Woldendorp, J.W., 1963, in *Mededelingen van de Landbouwhogeschool te Wageningen*, Nederland, **63**, 1–100.)

Soil type	Herbage	Roots	Soil	Not accounted for = loss by denitrification
Sandy soil	65%	10%	10%	15%
Clay	57%	11%	7%	25%
Peat (80% water content)	57%	2%	22%	19%
Peat (95% water content)	44%	2%	17%	37%

The activity of saprotrophic microbes around roots may, on balance, be slightly harmful. They affect root morphology, probably adversely, and their effects on nutrient availability and uptake are often detrimental or only marginally beneficial. If this is the case then to look at it teliologically, the plant is wasting photosynthate on all the root exudates, etc. It may be that there is some benefit which we do not see at present or perhaps the exudates and sloughed-off cells are all an unfortunate accident of the way the root functions in the absorption of water and nutrients. If the plant is nitrogen or phosphorus limited, as it may well be in most natural ecosystems, photosynthate is probably more than sufficient for both and it may not matter so much if some is lost through the roots provided this produces a marginal increase in phosphorus or nitrogen uptake. Most of these data on the effects of saprotrophs on nutrient uptake are derived from experimental situations which did not include predation on the bacteria and fungi by protozoa and other soil animals. Predation usually increases nutrient release. We just do not know enough about the microbiology of the root surface and the surrounding soil to answer these questions, or to manipulate the rhizosphere saprotrophs to the plant's advantage in agricultural situations.

Root-associated mutualistic biotrophs

There are two main associations, the nitrogen-fixing root nodules (Sprent 1979; Postgate, 1982) and the various mycorrhizal associations (Mosse *et al.*, 1981; Harley & Smith, 1983). There are other, apparently minor, mutualistic associations such as Cyanobacteria in roots of the angiosperm *Gunnera* and in cycads which will not be considered here.

Root nodules

Root nodules (Fig. 6.10) are associations between the plant root and the bacteria (including some actinomycetes) which fix atmospheric nitrogen, some of which is usually released to the host plant. The nodule bacteria are

Fig. 6.10 Root nodules on a legume that has been artifically inoculated with *Rhizobium*. (Photograph courtesy Karl Ritz, Department of Botany, Universty of Bristol.)

genetically capable of all stages in nitrogen fixation, but in the mutualistic associations they normally depend on the host for the considerable energy and reducing power requirement of the process and for the maintenance of a low oxygen tension in the nodule environment. Nodules formed by *Rhizobium* on legumes have been more intensively studied than those formed by *Frankia*. The latter occur in various trees and woody shrubs and have been studied most on *Alnus, Myrica* and *Casuarina*. The infection process seems to be similar to *Rhizobium* which will be described in some detail.

Nearly 13 000 species of legumes have been described and about 10% have been examined for root nodules. Of these almost 87% did in fact have nodules. There should still be about 10 000 host-*Rhizobium* associations to be found, described and understood! There is some degree of specificity between host and symbiont and different species of *Rhizobium* have been described on the basis of their host specificity (Table 6.12). This is however only part of the

Table 6.12 Cross-inoculation groups of *Rhizobium*. (From Postgate, 1982.)

	Organism	**Examples of host plants**
Fast-growing, acid-forming types		
Pea group	*R. leguminosarum*	Peas, broad beans, lentils, vetch
Bean group	*R. phaseoli*	Kidney beans, mung beans, runner beans
Clover group	*R. trifolii*	Clover
Alfalfa group	*R. meliloti*	Lucerne, melilot, fenugreek
Slow-growing types		
Lupin group	*R. lupini*	Lupins, seradella
Soybean group	*R. japonicum*	Soybeans
Cowpea group	'Cowpea miscellany'	Cowpeas, peanuts etc.

problem because minor changes in host cultivar or *Rhizobium* strain can greatly affect the amount of nitrogen fixed. It is also true, unfortunately, that efficient colonization and nodulation does not necessarily lead to nitrogen fixation: there can be nodules which do not fix nitrogen. The formation of legume nodules, in a few agriculturally important crops anyway, is now well understood and is summarized in Table 6.13. The rhizobia occur in the soil and some legumes specifically stimulate the correct strain to multiply; *Pisum sativum* exudes homoserine which preferentially stimulates growth of *R. leguminosarum*. The bacteria become attached to the root, especially the root hairs or they may be embedded in the mucilage. This stage has been suggested as one of the possibilities for specificity. Legumes produce specific proteins (lectins) from their roots and these may bind the bacteria by joining specific polysaccharide on the root hair to antigenically similar ones on the bacterium surface. In the presence of attached bacteria, or even high numbers of free bacteria in the soil, the root hairs curl and may become branched. This was thought to be due to indole acetic acid produced by the *Rhizobium* from tryptophan in root exudates, but it is now clear that the situation is more

Table 6.13 Host-rhizobia interactions during legume nodule formation. (From Sprent, 1979.)

'Developmental' stage	Genetical/physiological requirements	Comments
Rhizobium multiplication in soil	Host secretions may stimulate or inhibit	Degree of specificity variable
Root hair curling and branching	Compatible host and *Rhizobium*	Degree of specificity variable
Attraction between compatible host-rhizobial cells	? Matching of crossreacting groups. ? Bridging by lectins	Found by some to be highly specific: others less so
Entry of bacteria	Dissolution or stretching of host cell wall	Temperature affects strain of *Rhizobium* entering and thread development. Plants may have genetically controlled antinodulating factors
Growth of infection thread	Matched development of rhizobia with host cell wall material	
Formation of nodules	Correct balance of growth factors from both partners	*Rhizobium* mutants known which cause development to stop at any stage. Correct matching of host and rhizobial genotypes essential
Formation of mature infected cells	Release of rhizobia from threads. Matched growth of rhizobia and host membranes	
Formation of bacteroids, development of nitrogenase and haemoglobin, etc.	Specific interactions involving both partners	Host may act by providing correct environment
Maintenance of bacteroid tissue	Correct interchange of materials between symbionts	Not well understood. Strongly conditioned by environment

Increasing specificity of interaction between host and bacterial genomes ↓

complex and there may be several root hair curling factors. The curled, infected hairs then produce an infection thread or tube which grows back down the inside of the root hair. The infection thread wall is of host origin but is not continuous with that of the root hair. Bacteria pass down the infection thread as it crosses the root cortex to initiate the nodule. Nodules may be determinate, i.e. grow to a given size and then stop as in soybean, or if the infection thread can keep growing within the nodule then a branched indeterminate nodule may be formed (e.g. in pea or clover). The rhizobia may produce hormones and the infection stimulates plant production of auxin and cytokinin. This causes a division of cells in the cortex. As rhizobia are released from the infection thread (Fig. 6.11) the cells of the root expand to form the nodule in which the infected cells are surrounded by cortex and vascular strands. The bacteria remain within membrane-bounded vesicles within the cell (Fig. 6.12): since the membrane is derived from the infolding of the host plasmalemma the bacteria are actually outside the cytoplasm. Rhizobia in young nodules divide until the cell is packed with bacteria which may be branched in X- or Y- shapes (Fig. 6.11). Effective nitrogen-fixing nodules are pinkish in colour because of

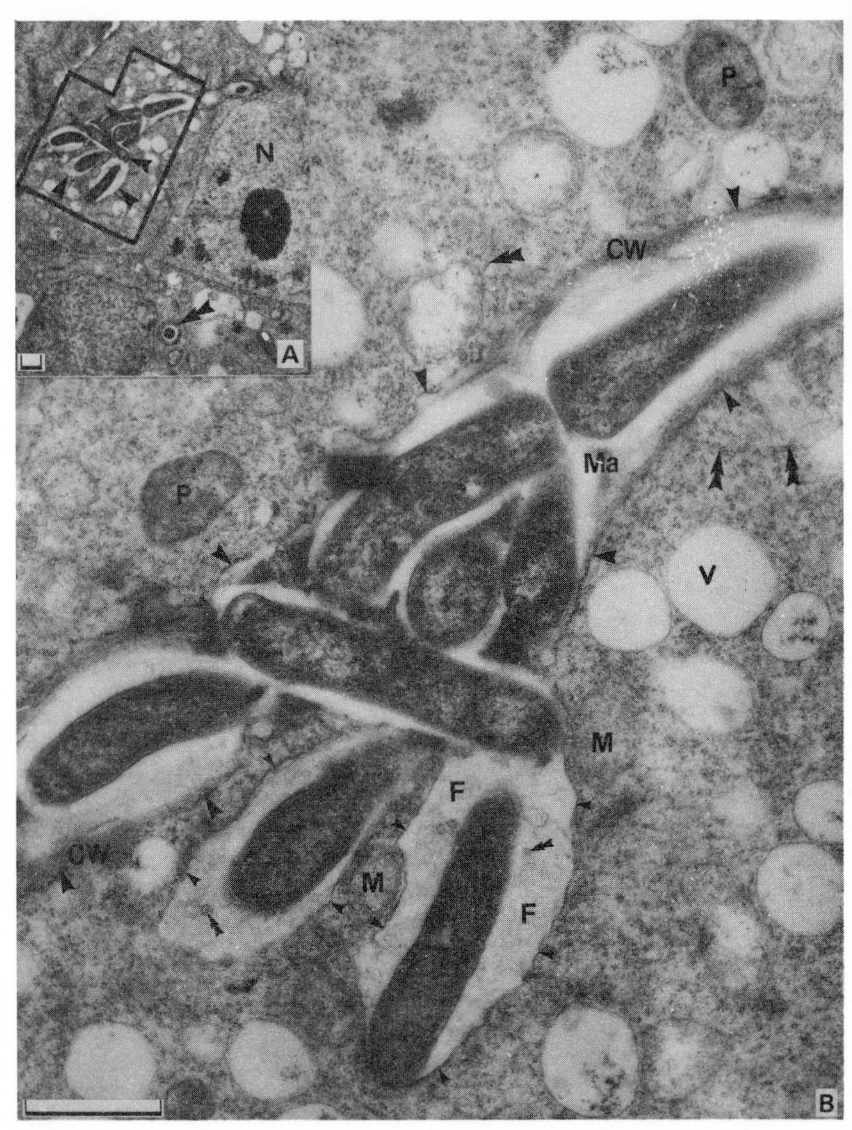

Fig. 6.11 **A.** Rhizobia are seen in the process of escaping (single large arrows) from an infection thread which traverses a host cell. Another infection thread cut in cross section (double large arrows) is shown. Bar = 1 μm. **B.** Higher magnification of the outlined portion of **A** showing three rhizobia in the process of escaping. A membrane (single small arrows), the future peribacteroid membrane, has formed bulges around these rhizobia and is continuous with the host plasma membrane (single large arrows adjacent to the infection thread cell wall (CW). Note the fibrillar material (F) and small vesicles (double small arrows) near two of the escaping rhizobia. Also shown are polyribosomes (double large arrows), proplastids (P), mitochondria (M), and small vacuoles (V) in the host cell. Bar = 1 μm. N = nucleus. Ma = thread matrix. (From Newcomb, W.F. McIntyre, L. 1981. Development of root nodules of mung bean (*Vigna radiata*): a reinvestigation of endocytosis. *Canadian Journal of Botany*, **59**, 2478–99.)

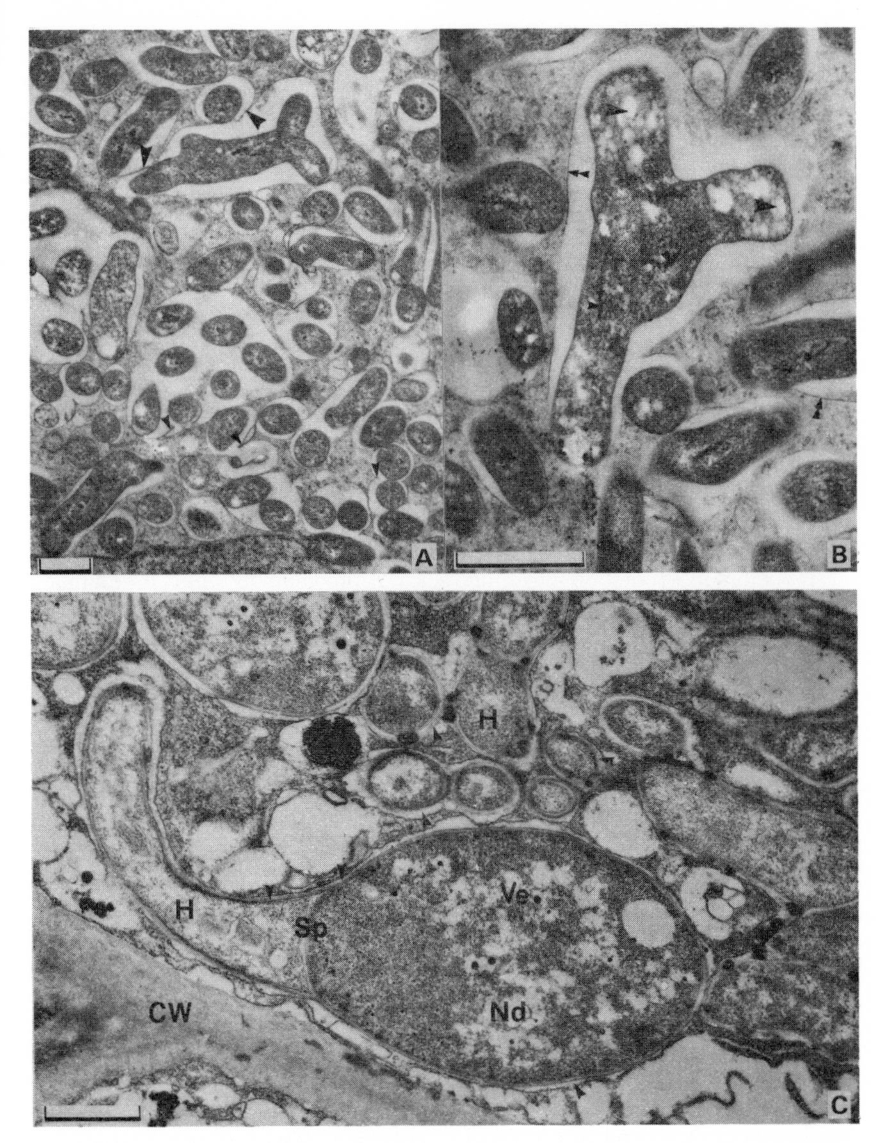

Fig. 6.12 **A.** A single large pleomorphic rhizobium, fusing (large arrows) peribacteroid membranes, and remnants (small arrows) of fused peribacteroid membranes are present in an infected cell. Bar = 1 μm. **B.** A bacteroid-like cell from a 17-day-old nodule contains internal membranes (single small arrows) and polyhydroxybutyrate deposits (large arrows). A peribacteroid membrane (double small arrows) surrounds the rhizobia. Bar = 1 μm. From Newcomb & McIntyre, see Fig. 6.11. **C.** *Frankia* in roots: the hyphal (H) connection to an egg-shaped endophytic vesicle (Ve). A septum (Sp) separates the vesicle from the hyphae. The vesicle contains more ribosomes than hyphae with the resulting different electron densities of these two cells. This is not always the case. Electron-dense inclusions are present within the nucleoid (Nd) regions of vesicles and in the capsule (arrows) adjacent to hyphae. CW × cell wall of host. Bar = 1μm. (From Newcomb, W. 1981. Fine structure of the root nodules of *Dryas drummondii* Richards (Rosaceae). *Canadian Journal of Botany*, **59**, 2500–14.)

the presence of leghaemoglobin which is thought to protect the oxygen-sensitive nitrogenase. After a few days or weeks the host cell cytoplasm degenerates and the rhizobia are no longer membrane-bounded. The nodule does not fix nitrogen after it has degenerated to this stage.

There is a very wide variety of legumes which are important in both the intensive western agriculture and in the third world. Protein from legumes is of vital importance to much of the world's population, either directly as in peas, beans, etc., or indirectly when used as forage crops (alfalfa, clover) for animal production. The conditions favouring fixation, in terms of temperature, soil moisture, etc., have been widely studied and have been shown to vary from host to host and with the *Rhizobium* strain. There is such a variety of legumes that there are species which are tolerant of drought, high temperature or other environmental extremes.

The nitrogen fixation has an energy demand of approximately 3.5 molecules of glucose per molecule of N_2 fixed. This is equivalent to 4 to 10 g of carbohydrate carbon per g of nitrogen gas. The bacteria use host photosynthate and this may amount to up to 10% of the host primary production. The alternative is to use nitrogen fertilizer, but nitrate has a similar energy demand (3.1 molecules of glucose per molecule of nitrogen assimilated) for it also must be reduced to $-NH_2$ before incorporation into amino acids. Ammonium fertilizers have a much lower demand (0.85 molecules of glucose per molecule of nitrogen assimilated). The situation is complicated by the effects of nitrogen source on leaf area and plant growth: for example nitrate-fed peas may have a growth rate 15–20% greater than nitrogen fixers.

As regards the actual amount of nitrogen fixed by nodules there is great variation according to the climate and other environmental conditions but some values are set out in Table 6.14. Legume–*Rhizobium* in general fix more than the *Frankia* systems but this does not necessarily mean that *Frankia* is less important, for most of the hosts are pioneer plants colonizing very poor sites where any added nitrogen would be important and, in general, forest crops do not have such a high nitrogen demand as agricultural legumes. Some of the values are very high, for example *Desmodium*, but this measurement was made under ideal conditions: generally in the tropics forage legumes do not do so well if there is a pronounced dry season or high temperatures. Apart from direct fixation by forest trees such as *Alnus* there is also interest in fixation by shrub understories or by crops such as lupins grown beneath the trees.

It is now possible in many countries to buy commercial *Rhizobium* inoculum to increase nitrogen-fixation either by introducing the appropriate strain where the crop had not previously been grown or to try to introduce a more effective strain into an existing situation. The latter requires that the inoculum can compete successfully with indigenous strains. Generally the rhizobia are grown in broth culture, inoculated into a peat and nutrient containing carrier prior to drying and packaging. This powdered peat may then be applied to seeds as a dust, slurry or pellet or it may be sprayed on the soil as the slurry or as granules. It is possible to get a very high yield increase (up to 100%) compared with un-inoculated controls, though different commercial inocula vary in their effectiveness, and conditions of storage can have a great effect. Once inoculation is accepted then there is a greater

Table 6.14 The quantity of nitrogen fixed by various symbiotic systems. (Based on Nutman P.S., 1977, *Symbiotic nitrogen fixation in plants*. Cambridge University Press, Cambridge; and on various data from Campbell R., 1983, *Microbial Ecology*, 2nd Edition, Blackwell Scientific Publications, Oxford.)

Bacterium	Host	Nitrogen fixed (kg N ha^{-1} y^{-1}) Mean	Range
Rhizobium	**Forage and browse legumes**		
	Clover (*Trifolium*)	183	45–673
	Lucerne (*Medicago*	—	128–300
	Centro (*Centrosema*)	259	126–395
	Tick Clover (*Desmodium*)	897	—
	Stylo (*Stylosanthes*)	124	34–220
	Acacia (*Acacia*)	270	—
	Pulse legumes		
	Field bean (*Vicia*)	210	45–552
	Pea (*Pisum*)	65	52–77
	Chick pea (*Cicer*)	103	—
	Ground nut (*Arachis*)	124	—
	Soybean (*Glycine*)	—	57–94
Frankia	Alder (*Alnus*)	—	56–150
	Myrica	9	—
	Sea buckthorn (*Hippophae*)	—	15–179
	Casurina	58	—
	Snowbush (*Ceanothus*)	60	—

possibility for the improvement of the *Rhizobium* strains, by genetic engineering as well as by simply selecting the best strains. Inoculation of the *Frankia* endophyte looks as though it would be useful but it is very difficult to grow in culture.

In contrast to the free-living microbes (p. 122) the effect of symbionts on the nitrogen economy of the plant can be dramatic, and definitely for the benefit of the plant, even though it uses some photosynthate to supply the bacteria with a carbon source the plant usually gains net benefit. Where fertilizers are available, even when they are expensive, the choice becomes more difficult and depends on the efficiency of the particular *Rhizobium*–plant association in relation to the energy costs and effectiveness of fertilizers. It seems certain however that quite good advances can be made in the field for improving symbiotic nitrogen fixation, especially in the third world.

Mycorrhizas

Mycorrhizas are mutualistic associations (p. 14) between fungi and plant roots. Such a general definition covers many different types of association which have entirely different relationships with the host. There are for example those with some ericaceous plants and with orchids where the fungus provides carbohydrate, derived from litter decomposition, to the higher plant which may be achlorophyllous. Such fungi are saprotrophs. These are of scientific interest (Mosse *et al.*, 1981; Harley & Smith, 1983), but not important in agriculture and forestry. There are, however, two sorts of mycorrhizas which are of overwhelming importance, the ectomycorrhizas of forest trees, especially temperate trees, and the vesicular arbuscular mycorrhizas of very many plant species except Cruciferae and some Chenopodiaceae, Cyperaceae and Caryophyllaceae.

Ectomycorrhizas are formed by basidiomycete fungi, especially the mushrooms and toadstools, and ascomycete associations are also known. The mycelium forms a sheath over the outside of the root which may take many forms from a loose weft of hyphae to pseudoparenchymatous tissue usually with hyphal extensions into the surrounding soil (Fig. 6.13B). These hyphae may exploit nutrient supplies outside the range of the unaltered host roots and they may join together roots of the same or different trees. Nutrients may pass from tree to tree via the mycorrhizal fungus and there is evidence, for example, that young trees growing in the shade of their parent may receive nutrients from that parent so that they can grow in the dry, nutrient deficient, shaded soil. The sheath around the root alters the growth pattern to produce characteristic short, thick branches on the roots (Fig. 6.13A). The hyphae penetrate between the epidermal and cortical cells but do not usually enter through the cell walls (Fig. 6.13C). The amount of hyphae within the root, forming the sheath, or in the soil will vary with the host tree and the fungus. Some fungi, which are always associated with a particular tree, show great host specificity; for example some *Russula* species occur only near beech trees (*Fagus* sp.). Other fungi. such as *Cenococcum*, have very catholic tastes and form morphologically dissimilar associations with many hosts. Various possible morphological forms have been classified for each host species. The sheath may be very substantial, accounting for up to 40% of the root weight and in addition the fungi may produce quite large fruiting bodies in the right season. It is normal for a single plant, especially a large tree, to form associations with more than one species of fungus. There are data on the occurrence of fruiting bodies which suggest that there could be successions of different fungal species forming mycorrhizas with a single host plant, the most complex mixtures of species occurring in older plants. The host species are very variable, classical examples are the northern temperate conifers and hardwoods such as pines, firs, spruces, Douglas fir, larch and beech, oak, birch, etc. There are however also quite a lot of tropical or sub-tropical species of trees, shrubs and herbs including dipterocarps and legumes that form ectomycorrhizal roots. Notice that it is possible to get mycorrhizal roots with nitrogen-fixing nodules (p. 147).

The exact physiological relationship between the ectotrophic fungus and the host has been studied in many examples and is assumed to apply to all ectomycorrhizas. The host is autotrophic and supplies the fungus with available carbon, most of the fungi forming such associations are not able to degrade the cellulose, lignin and other carbohydrates of the leaf litter in which they grow. Plants exposed to $^{14}CO_2$ translocate ^{14}C-sugars to their roots and label appears in the fungal sheath; much more carbon (three or four times as much) is translocated to mycorrhizal roots as goes to non–mycorrhizal roots. The whole mycorrhiza stores glucose and fructose while only the root contains sucrose and starch and the fungus contains trehalose, mannitol and glycogen.

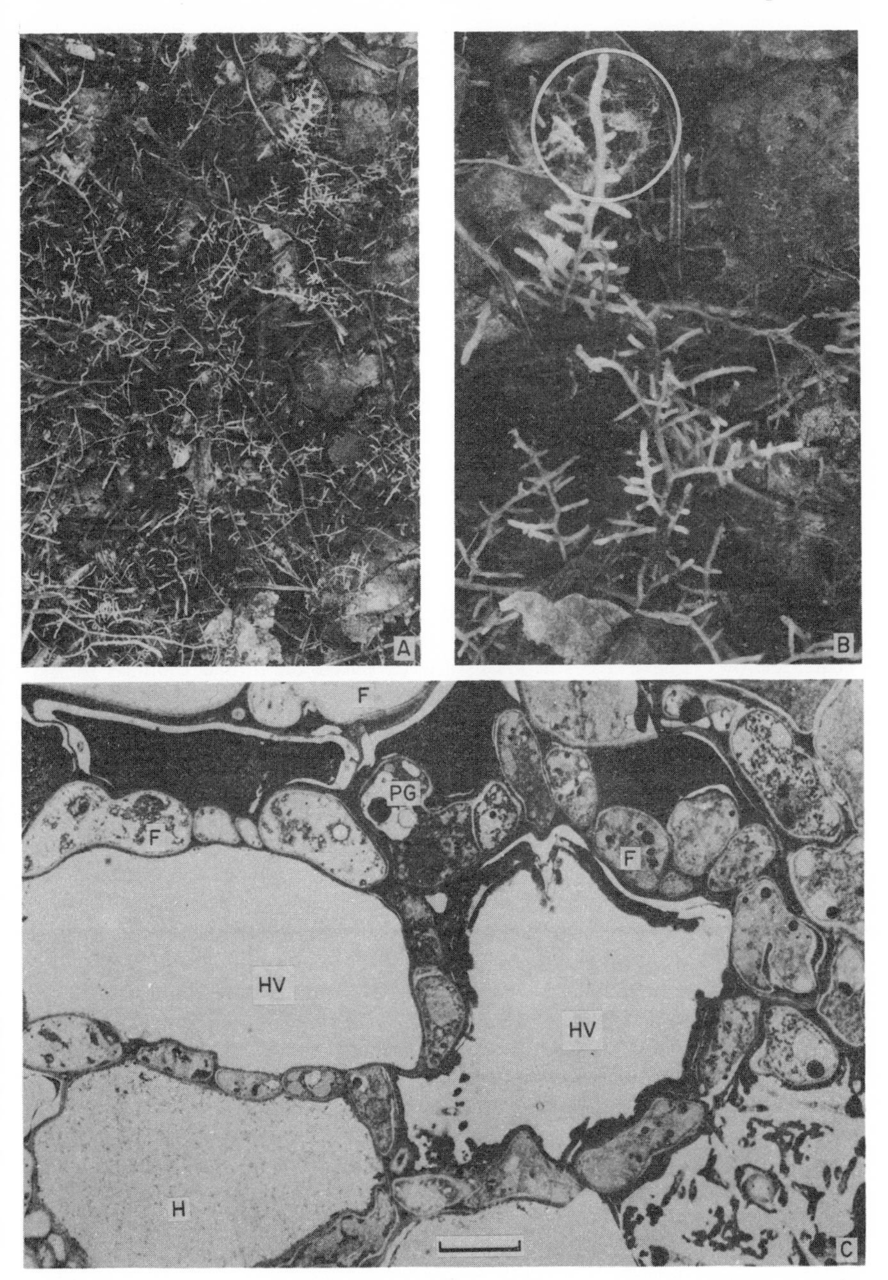

Fig. 6.13 A. Beech leaf litter with the surface leaves and debris removed to show the mycorrhizas growing on the surface of the mineral soil. **B.** Detail of A to show the hyphae growing away from the sheath (circled). **C.** Transmission electron micrograph of ectomycorrhizas of *Corylus avellana* with *Tuber macrosporium* showing the sheath at the right and the Hartig net extending between the cells. Bar = 5 μm. F = fungus; PG = polyphosphate granule; HV = host vacuole; H = host cell. (From Scannerini, S. & Bonfante-Fasolo, P., 1983, *Canadian Journal of Botany*, **61**, 917–43.)

The ectomycorrhizal fungus is thus a drain on the host's photosynthate, though how it does this is not clear: various suggestions have been made including the production of specific substances to make the host cell 'leak' and the simple formation of a diffusion sink by the fungus so as to maintain the flow of nutrients from the root.

The main effect of the mycorrhizal association on the plant has long been known to be an increase in plant growth, and phosphorus uptake has been especially studied (Table 6.15), though other nutrients are also affected. There is not invariably an increase in phosphorus uptake (see third set of data, Table 6.15) and there is no evidence that novel or normally unavailable sources of phosphorus are used. Phosphorus is a very immobile element in the soil so any root normally uses available phosphorus from around itself and, as this is not rapidly replaced, a zone of depletion is produced around the root. The external hyphae increase the volume of soil which can be so exploited. The phosphorus may accumulate in the sheath, partly as polyphosphate, especially when soil phosphorus levels are low. Other substances such as nitrogen and potassium also accumulate in the sheath, either by active absorption or by passive diffusion in or through the sheath.

So the host plant has significant improvement in growth and availability of some nutrients, especially in those soils that are nutrient deficient. The price paid by the plant is in photosynthate supplied to the fungus and this may be considerable. There are various estimates based on fungal biomass in the association, production of fruiting bodies, respiration rates, etc., but all agree that the amounts involved are significant and may be 10–15% or more of the net photosynthesis. This drain may not be so serious as it at first appears because if the plant's growth is anyway nitrogen or phosphorus limited then there may be excess carbohydrate (see also p. 124). Whatever the real or apparent cost to the plant, the association is overall beneficial since the plant grows more and has improved concentrations of phosphorus and nitrogen.

The vesicular arbuscular mycorrhizas (VAM), are very different in their morphology and host relationships, but remarkably similar in the effects that they cause. The fungi involved are traditionally considered to be aseptate phycomycetes which used to be placed in the genus *Endogone*. This has now been divided into several genera (e.g. *Glomus*, *Acaulospora*, *Gigaspora*) largely on the basis of the spore morphology. They seem to be obligate biotrophs (p. 23) and in contrast to the ectomycorrhizas are very diffficult or impossible to grow in culture and this considerably complicates their study and commercial exploitation. Various other fungi have been reported to form VAM, the most common of which is the so-called 'fine endophyte'. There are many different host plants, from bryophytes and ferns to angiosperms and gymnosperms in both temperate and tropical regions of the world. In some angiosperm families VAM are uncommon and even some particular genera (e.g. *Lupinus*) within otherwise heavily infected families (Leguminoseae) seem not to have extensive mycorrhizal formation. It is therefore possible to have multiple symbiotic associations with VAM and root nodules (p. 147). There may even be VAM with ectotrophic mycorrhizas on the same plant. VAM are so important because their host range is so wide that they affect almost all natural communities and many agricultural crops throughout the

Table 6.15 Growth and specific nutrient uptake of nitrogen, phosphorus and potassium by *Pinus strobus* seedlings in three separate experiments with and without various degrees of mycorrhizal infection. Figures in [ ] are mean per cent increase (+) or decrease (−) in the mycorrhizal value in relation to the uninfected or lesser infected control. SA is the specific absorption in mg mg^{-1} root dry weight. (Modified from data assembed by Harley & Smith, 1983.)

Degree of mycorrhizal infection	Plant dry weight (mg)	Root weight (mg)	Nitrogen (SA)	Phosphorus (SA)	Potassium (SA)
Mycorrhizal	448	180	0.030	0.0047	0.019
	361	170	0.027	0.0042	0.015
	[+26]	[+3]	[+81]	[+225]	[+105]
Uninfected	300	174	0.013	0.0013	0.006
	361	182	0.018	0.0015	0.011
	301	152	0.016	0.0014	0.008
Mycorrhizal (70.2%)	337	127	0.042	0.006	0.016
	[+86]	[+69]	[+45]	[+200]	[+45]
Uninfected (9.5%)	181	75	0.029	0.002	0.011
Mycorrhizal (87%)	223	120	0.017	0.003	0.013
	[+140]	[+114]	[−19]	[−25]	[−28]
Uninfected (10%)	93	56	0.021	0.004	0.018

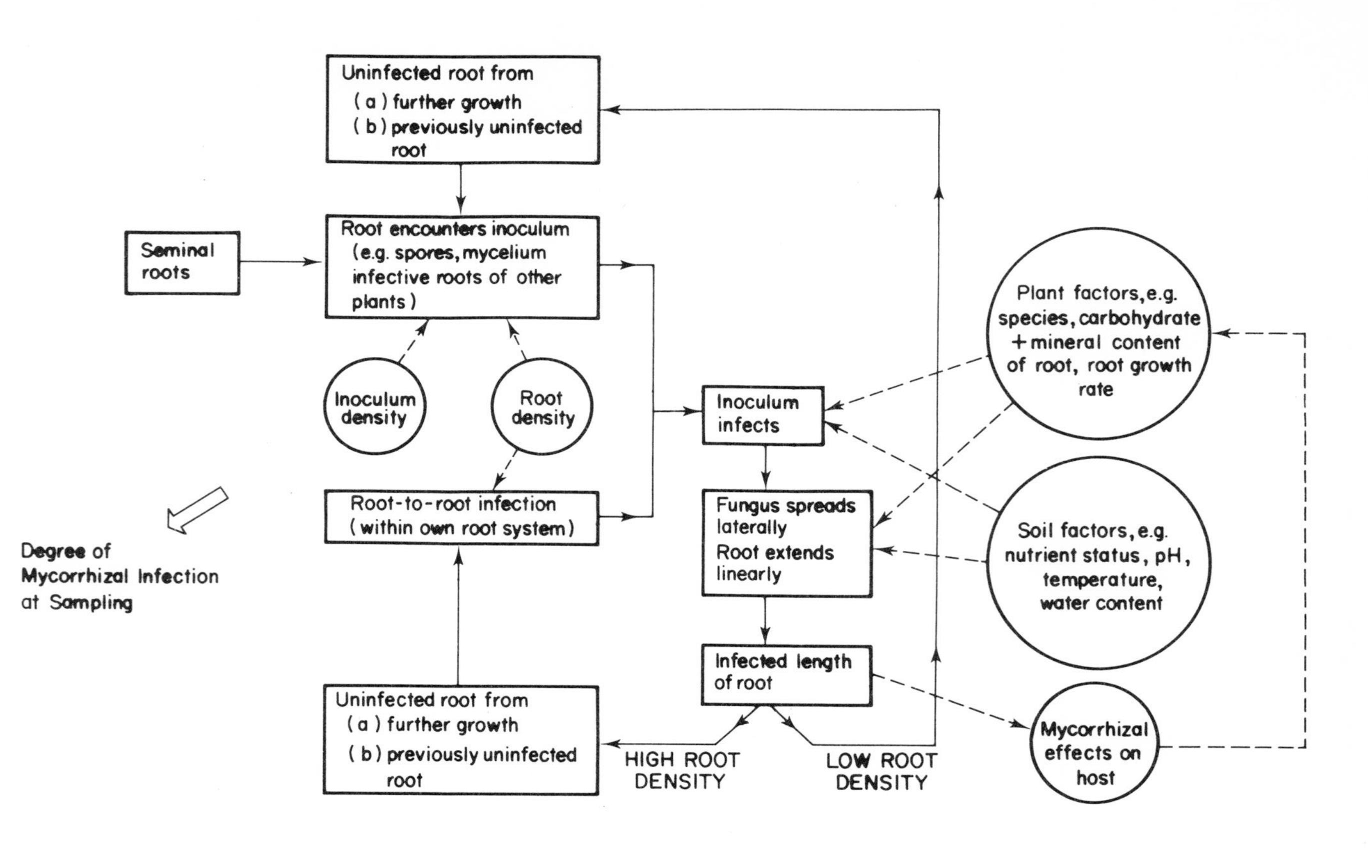

Fig. 6.14 Factors that may control the progress of mycorrhizal infection in a developing root system. (From Mosse *et al.*, 1981.)

world. They are the main form of mycorrhiza in most of the tropics and occur on many economically-important third world crops (rubber, cassava, cocoa, citrus, oil palm, etc).

The germination and growth of VAM fungi through the soil is affected by the pH, aeration and nutrients (see Mosse *et al.*, 1981; Fig. 6.14). There is also the complication of various root effects such as exudates and the presence of already infected roots. Obviously if the root density is high and inoculum potential of the fungus is also high then infection is more likely. Assuming these various factors are all favourable a hypha from the fungus contacts and penetrates the root, growing down into the cortex (Fig. 6.15A) and along the root surface. Extracellular hyphae may expand into vesicles between the cells of the inner cortex (Fig. 6.15B). Arbuscules are also formed in the cortical cells (Fig. 6.15B) by penetrating the wall, though not the plasmalemma which surrounds all the many branches of the fungus. There is a matrix of unknown origin and composition between the host plasmalemma and the fungal wall. This is very similar to the structure of haustoria of obligate biotrophic parasites (p. 27). The arbuscule almost fills the cell though the host cytoplasm remains and is rich in organelles some of which may be altered from the normal state; for example plastids may become chromoplasts (Fig. 6.16). After a few days or at most a week or two the arbuscule becomes vacuolated and the empty areas are walled off and collapse.

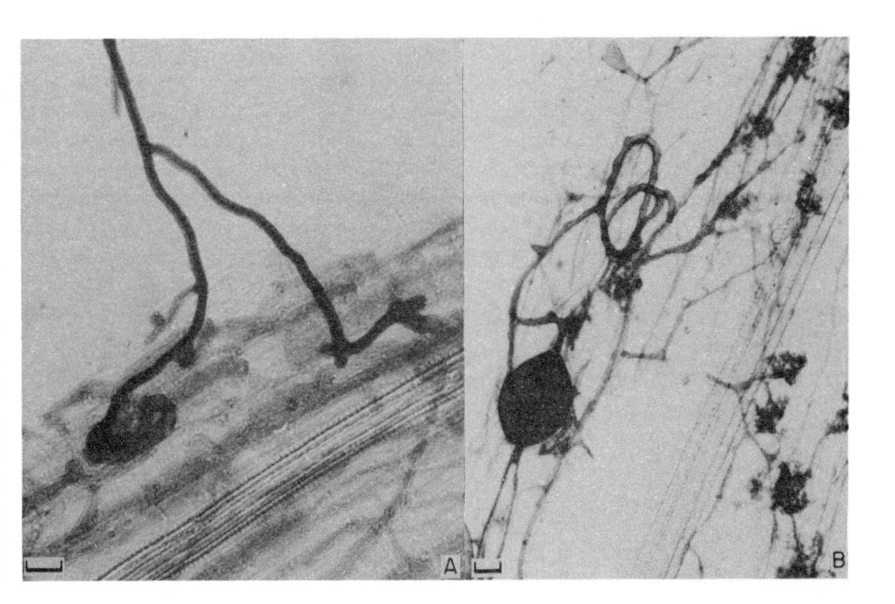

Fig. 6.15 A. The penetration of a root cell by a hypha of a VAM fungus. Bar = 10 μm. **B.** Vesicles and arbuscules within cortical cells. Bar = 10 μm. (Photographs courtesy of Karl Ritz, Department of Botany, University of Bristol.)

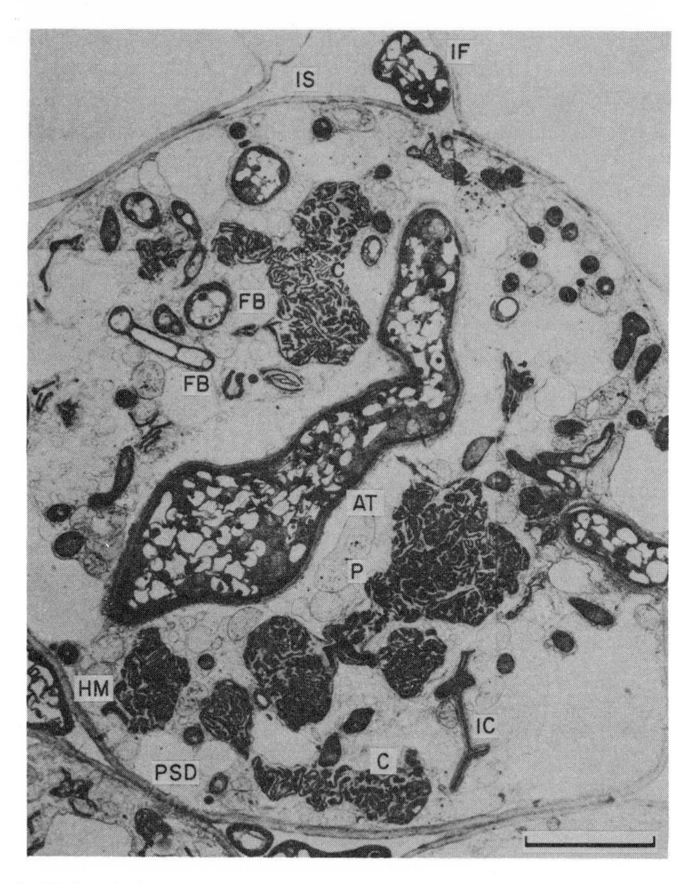

Fig. 6.16 Transmission electron micrograph of VAM of *Ornithogalum umbellatum* infected with *Glomus fasciculatus*. Cortical cell of root with the endophyte in the arbuscular phase. All parts of the arbuscule are shown, including arbuscule trunk (AT), fine branches (FB) and isolated clumps (IC), and the infection hypha (IF) is visible in the intercellular space (IS) at the top of the picture. Host plastids (P), plasmodesmata (PSD) and mitochondria (HM) are normal, though the chromoplast (C) is abnormal. Bar = 5 μm. (From Scannerini, S. & Bonfante-Fasolo, P., 1983, *Canadian Journal of Botany*, **61**, 917–43.)

It seems that the physiological relationship between the host and its VAM fungi is similar to the ectotrophic ones already described. The fungus receives carbohydrates from the host but there does not appear to be any conversion to trehalose or mannitol. There are no very good estimates of the amounts involved but quantities as low as 1% of the total photosynthate are quoted. This difficulty is largely because of the impossibility of separating the fungus from the host cells satisfactorily whereas in the case of ectomycorrhizal fungi the sheath can be removed relatively easily. There is usually an increase in plant growth in those plants infected by VAM especially on soils low in phosphorus or on heavily nitrogen-fertilized soils where phosphorus is made limiting by the excess of nitrogen. As well as more growth there may be increased nutrient levels in the plant tissues, especially increased phosphorus. As with ectotrophic mycorrhizas this is not thought to be due to the exploitation of any exceptional sources but to the increase in the volume of soil exploited by the external hyphae. Phosphorus has been shown to be translocated along the hyphae towards the root and it accumulates inside the root in the arbuscules as organic polyphosphates. It then reaches the host plant but it is not clear whether this is across the active arbuscule surface or when the arbuscule dies and collapses in the host cell.

VAM are, therefore, very important in the growth of plants, especially in phosphorus limited soils which are common in both agricultural and natural ecosystems all over the world. Many tropical soils are particularly phosphorus deficient. The very fact that so many plants of so many different genera have VAM confirms their importance and as we shall see there are many interactions with saprotrophic and parasitic organisms in the soil, as well as the direct effects which VAM have on their hosts.

Parasitic root-infecting fungi

Root pathogens have to compete with the normal soil and root flora, and have to survive in the absence of the host plant. They must therefore either be not too particular as regards their host so that a suitable plant is usually available, or they may have dormant spores or sclerotia, or survive saprotrophically in general soil organic matter or on the remains of their dead host. The latter requires that they have a good competitive ability (Garrett, 1970) as well as being pathogenic.

There are some diseases which can affect roots of seedlings such as damping-off (p. 47). There are also a very large number of minor pathogens which are thought to cause rather vague symptoms of loss of vigour, some root death, differences in growth rates, etc. Such effects come to be recognized when control measures, such as fungicide application, are used even in the absence of any specific disease. Crops may then show growth or yield increases which are sometimes attributable to the control of these minor pathogen complexes. Such effects are difficult to quantify and it is therefore difficult to justify research. However these problems, along with serious root diseases, are now recognized as important. Major foliar diseases are controlled by fungicides and plant breeding, so the damage to roots becomes the next problem to solve.

A number of foliar or stem diseases, e.g. eyespot of cereals (*Pseudocercosporella herpotrichoides*) which overwinter in soil on roots or on rubbish left from the previous crop (called trash-borne diseases) will not be considered here. Root diseases themselves can be divided into those caused by: (1) some distorting or gall-forming organisms, (2) viruses, (3) vascular wilts, or (4) some form of rotting.

Distortions of root growth

One of the most widely described and studied of these diseases is club root of brassica crops, caused by *Plasmodiophora brassicae*. This organism has motile zoospores which infect the root hairs by first forming an adhesive pad and then penetrating the walls with a special hardened organelle called a stylet which is pushed through the cell wall. The amoeboid stage is then injected into the cell and divides and grows until after a few days it forms zoospores within the cell. These secondary zoospores are released into the soil and may there fuse to give a binucleate zoospore which again penetrates the root and will develop into a multinucleate mass of protoplasm (a plasmodium). These grow and divide, and by increasing root hormone levels such as indole acetic acid or indoleacetonitrile cause division and growth of root cells to form large swollen regions in which resting spores are formed. These spores can persist in the soil for many years before germinating and infecting root hairs again. This rather unusual organism is related to myxomycetes and is not considered a true fungus by many authorities, though it has traditionally been studied by mycologists. There is no way to kill the organism in the soil except complete sterilization which is not usually practical. Roots can be protected from zoospore penetration by various root dips at transplanting time (e.g. calomel, mercurous chloride) and some varieties of brassicas are less affected than others. Most plants can tolerate mild infection but in severe cases uptake of water and nutrients is reduced and the plants wilt easily and do not grow well.

The only other common distortions of root growth are caused by nematode cysts (e.g. cereal root cyst-nematode, *Heperodera aveni*, or cotton root knot, *Melodogyne incognita*) which are outside our present discussion, but may be confused with damage by micro-organisms. Similarly various nematodes cause root rots or lesions which may appear similar to fungal lesions, especially when the nematode damage is later invaded by saprotrophic fungi. Nematodes also transmit a number of soil-borne viruses to the roots of plants (see p. 4, e.g. ringspot virus, nepovirus). Some soil fungi, especially those related to the club root fungus, also transmit viruses. *Polymyxa graminis* and *Spongospora subterranea* transmit, in their zoospores, wheat mosaic and potato mop-top virus respectively.

Virus diseases

Viruses do occur in roots, except perhaps in the apical meristem, as indeed they must if some are soil or root-aphid transmitted and symptoms appear in the leaves. However virus diseases of roots seem to be uncommon; is this a genuine effect or do virologists spend all their time looking at the leaves and not at the roots? There may of course be a loss of root vigour when the top is severely infected but actual roots lesions are rare; tobacco etch virus produces root necrosis on the pepper cultivar Tabisco (*Capsicum frutescens*).

Vascular wilt diseases of angiosperms

These are mostly caused by fungi, which initially anyway, infect the roots. There are also some bacterial infections that produce similar symptoms. Many different organisms are involved on different hosts but the fungi *Fusarium* and *Verticillium* are very important. *Ceratocystis ulmi*, causing Dutch Elm disease, has recently become notorious and is a wilt which is carried by air-borne insects rather than infecting through the roots (p. 103). The bacteria

Pseudomonas solanacearum and *Corynebacterium michiganense* both cause vascular wilts and may enter through root wounds caused by animal feeding or other means.

The organisms enter the vascular system, especially the vessels, either by direct wall penetration or via tyloses (Fig. 6.17C) and damaged pits in the case of bacteria. The organisms proliferate in the conducting vessels and may cause discolouration of the xylem (Fig. 6.17D), chlorosis of leaves, wilting (either transitory or permanent), etiolation, epinasty (downward drooping of leaf stalks; Fig. 6.17B) and finally death and defoliation of the parts of the plant served by the infected areas of the xylem (Fig. 6.17A). Generally only a small part of the xylem is at first infected and spread between vessels is difficult for bacteria which depend upon holes and, even though fungi can penetrate walls, they usually pass through the pits. The vessels colonized are usually no longer functional because of mechanical blockage, but the root and stem xylem as a whole is still functional. What then causes the wilting?

Wilt pathogens produce chemicals which have been called toxins. Bacterial 'toxins' are high molecular weight polysaccharides or glycopeptides which act by blocking the xylem elements especially the pits, in the stem and leaf petiole where the small number of vascular strands makes the water flow particularly sensitive to vascular disruption. These compounds are not therefore true toxins as they act by mechanical blocking of water flow rather than by chemical poisoning.

Fungi causing vascular wilts produce a variety of extracellular compounds which have been implicated in symptom production. There are polysaccharides which, as in bacteria, may cause vessel blockage. There are also water soluble, low molecular weight compounds such as fusaric acid (5-butylpicolinic acid) and lycomarasmin (N–(hydroxypropionic acid)–glycyl asparagine) which are produced by *Fusarium oxysporum* f. sp. *lycopersici* on tomato. Fusaric acid produces some, at least, of the wilt symptoms in the absence of the fungus and has been found in infected plants: it is probably a true toxin. The case for lycomarasmin is less clear, for although it can produce symptoms at high concentrations they do not closely mimic the diseased condition and lycomarasmin has not been found in diseased plants. Hydrolytic, wall-degrading, enzymes produced by the fungi have also been implicated in the wilt syndrome by causing vessel blockage with degradation products and by attacking the host cells direct. Increases in auxin (IAA) and ethylene concentrations also occur and may be responsible for the symptoms of epinasty and etiolation.

The wilt syndrome is therefore the result of a very complex interaction between the host and the pathogen which in many cases is poorly understood: they are none the less very important diseases in agriculture and forestry throughout the world.

Root rots

There are remarkably few bacterial rots of growing roots and none that are important on a world scale. There are: (1) opportunistic infections of wounded or weakened plants, (2) several important wilt diseases (discussed above)

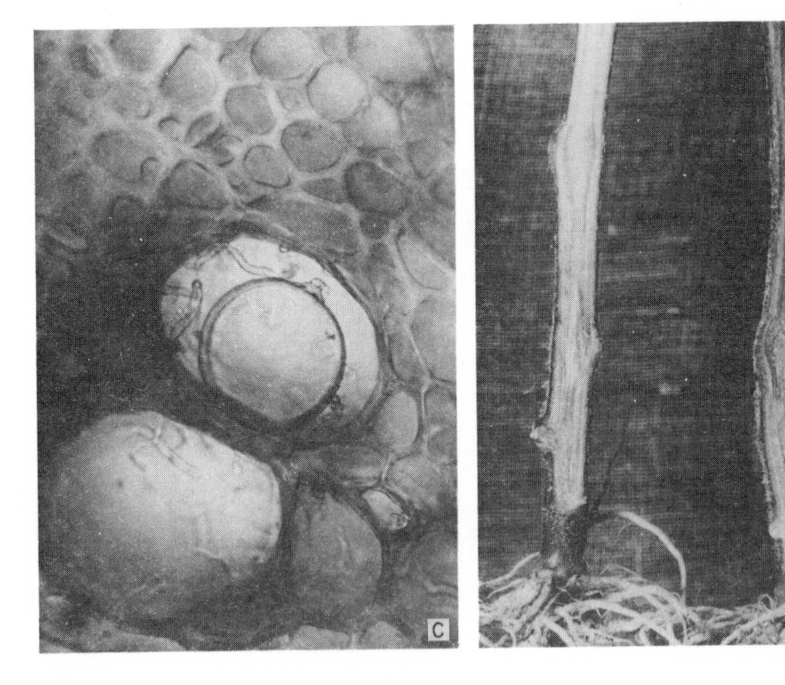

Fig. 6.17 **A.** Lucerne infected with *Verticillium albo-atrum* which shows complete death of the plant. **B.** Tomato infected with *V. albo-atrum* showing etiolation and epinasty. **C.** Tomato infected with *V. albo-atrum* showing the vessels containing tyloses and hyphae. **D.** Tomato infected with *V. albo-atrum* showing vascular browning. (Photographs courtesy Julie Flood, Department of Botany, University of Bristol.)

which may end with decay of the plant, (3) diseases of the stem base (e.g. blackleg of potato, caused by *Erwinia carotovora* var. *atroseptica*) which may spread to the roots and (4) there are very important soft rots of root storage organs which are considered later (p. 145) as post-harvest decays.

In contrast fungal infections of growing roots are very numerous and produce many important necrotic diseases. There are many unspecialized necrotrophs which attack weakened roots or young tissue causing seedling diseases (p. 47) or secondary infections of unhealthy plants. There are unspecialized pathogens (facultative or obligate necrotrophs, p. 22) that will attack apparently healthy plants. There is most damage from such organisms when the plants are stressed by water shortage, disturbance, etc., or when the pathogen is an introduced alien as in the case of *Phytophthora cinnamomi* in Western Australia, which is causing severe disease in eucalyptus. *P. cinnamomi* also attacks other horticultural crops such as avocado (*Persica americana*) and pineapple (*Ananas comosus*). It has a very wide host range of well over 400 species, particularly woody dicotyledenous shrubs and trees. The obvious symptoms are chlorosis, dyeback of shoots and eventually death of the whole shrub or tree. This is caused by death of roots, usually small feeding roots, but occasionally roots up to 10 cm diameter are attacked to produce a soft, spongy rot. This disease is very difficult to control, since fungicides are uneconomic on a forest scale and anyway it is not possible to treat established tree roots in the soil by drenching with fungicide. Biological control has been tried with some possibility of success (p. 150).

Fusarium species are also unspecialized pathogens and are common inhabitants of roots causing various root rots, for example of French bean (*Phaseolus* sp.). They have recently received most attention for damage they do to cereals: *F. culmorum* and *F. avenaceum* (= *Gibberella avenaceum*, the perfect stage) are necrotrophs which cause seedling blight, foot and root rots and head blight throughout the world. They are soil-borne or trash-borne organisms with a high competitive ability. The attack is usually most noticeable in the stem bases, but roots are also affected.

A rather more specialized necrotroph affecting wheat roots is *Gaeumannomyces graminis*, the take-all fungus. It may cause seedling death or the plants may become infected later; such plants produce poor seed heads which ripen prematurely as 'white-heads'. Roots are killed and blackened and plants only survive by producing nodal roots faster than the fungus kills them. The fungus is considered to be a specialized necrotroph because it grows ectotrophically on the roots by means of black runner-hyphae. These grow up the root surface, sending down smaller hyphae into the cortex and eventually to the stele. These small feeder hyphae kill the root cells but any host reaction that they provoke is not serious because the runner hyphae have already passed on up or down the root and started a new infection and such multiple infections rapidly overcome host resistance, and allow the fungus to invade the whole root. Again there are no effective fungicides or in this case any resistant varieties of wheat, and control depends upon cultural practices and conscious or unconscious use of biological methods (p. 150).

Such ectotrophic growth also occurs with other pathogens such as *Heterobasidion annosum* (= *Fomes annosus*) and *Armillaria mellea* (Garrett,

1970) which both attack tree roots and illustrate several other features of root pathogens. *Heterobasidion* grows on the root surface, especially on pine trees in alkaline soils, and *Armillaria* has specialized mycelial aggregations called rhizomorphs which grow for considerable distances through soil from an existing stump or root which acts as a food base. The rhizomorph, though composed of hyphae, has an apical meristem, a protective covering of melanized hyphae and conducting hyphae in the centre. On contacting a new root the bark is penetrated and the rhizomorphs grow up the cambium to the base of the stem and even up the stem itself, killing and colonizing the underlying wood as it goes. This results in dead roots and eventually rotten wood. Small shrubs which are attacked die very quickly and even in large trees the first symptom may be wilting, and death of leaves as roots are killed long before they have rotted enough for the tree to fall. Such dead shrubs and trees usually have mycelial fans or rhizomorphs or both underneath the bark.

Heterobasidion is interesting because it is very important in timber decay (p. 102) and also it shows how an understanding of the life cycle can lead to control of the pathogen. Despite the fact that it causes a stem rot, it can be considered here too, as a root pathogen, because it spreads from tree to tree by root contact and natural root grafts in dense forest. The primary infection of the healthy site is, however, by basidiospores landing on wounded stems or cut stumps (e.g. in thinning plantations). The fungus grows down into that stump before attacking nearby trees from underground. Prevention of initial infection is crucial since once in the roots it is impossible to reach the fungus. *Heterobasidion* does not have a high competitive ability. Protection may be by treating the stump with fungicide such as creosote, or by applying nutrients such as urea which encourage saprotrophic fungi to keep out *Heterobasidion* or in some cases applying saprotrophs direct as spores (e.g. spores of *Peniophora* (= *Phlebia*) *gigantea*). All these methods to varying extents are biological control as they exclude the pathogen until saprotrophs are established, encourage saprotrophs or actually inoculate the surface with them. Once the roots are colonized, *Heterobasidion* cannot replace the resident population.

It must be obvious by now that control of root diseases in general is difficult. Fungicides, even the present commercial systemic ones, are not downward translocated, so unless the plant can be dipped in fungicide, during transplanting, there is no effective way of getting the chemical to the roots. Seed dressings will work for very young plants but do not persist. Resistant cultivars of plants are a long-term hope and a manipulation of environmental factors, via agricultural practices, or the other organisms in the soil, offer some solutions to the problem. It is also difficult to work on root diseases, quite simply because they are not visible until the plant is dug up. Assessment is not a relatively simple matter of counting leaf lesions at various stages, it may be necessary to destroy the crop to see if it is diseased at all or to find out how severe the disease is. However, as noted previously (p. 139) the root diseases are now getting more attention from plant pathologists. Finally in this section it should be stressed that only a minute proportion of root diseases have been discussed, merely picking out a few examples to illustrate some general principles.

Post-harvest decay of root crops

Such decay may be caused either by necrotrophic bacteria or fungi. The main bacterial problem is soft-rots, caused by *Erwinia carotovora* especially in carrots, as its name suggests, though it attacks a wide variety of non-root vegetables (e.g. potatoes) and fruits. The bacterium produces pectolytic enzymes which macerate the tissue. Preventive measures include storage at low temperature and particularly the prevention of damage to the root during lifting, processing and storage. Some biocides can be used as dips or in the wash water during the processing of vegetables.

Though bacteria are the predominant organisms on freshly harvested root crops there are many fungi that can cause rots in storage including those associated with field decays. Most are common organisms of soil or plant debris such as *Botrytis*, *Rhizoctonia*, *Alternaria* and some *Rhizopus* species. There must usually be some mechanical damage to allow entry, though physiological damage because of poor storage conditions also provides possible entry points.

Interactions between micro-organisms on or in roots

Mutualistic symbionts

Mycorrhizas have an effect on the saprotrophic root microflora and on other symbionts. Ectotrophic and vesicular arbuscular mycorrhizas have their own 'rhizosphere' which may show greater stimulation of bacteria than around the non-mycorrhizal root.

With an established fungal sheath in ectomycorrhizas all the root exudates pass through the hyphae which extract some components and add exudates of their own, so that what emerges from the mycorrhiza is different from the non-mycorrhizal root of the same plant. The microbial populations may be stimulated in both situations, and many have bacteria and fungi in common but there are also organisms unique to the mycorrhizal rhizosphere. Some of these may stimulate or enhance the infection of the root by the mycorrhizal fungus.

Saprotrophic bacteria may depress colonization by ectotrophic mycorrhizas, enhance it or have no effect. The outcome depends on the particular species of fungus and bacterium (Table 6.16) so that some fungi while generally good colonizers and able to compete with bacteria, are harmed by particular strains, or conversely some bacteria which promote growth of most of the fungi tested are antagonistic to one particular fungus (e.g. *Bacillus* sp. 2 and *Pisolithus tinctorius*). The situation in the soil must be very complex with competition or synergism between all the organisms in the soil and on the root and the mycorrhizal antagonists being themselves out-competed.

A further complication of these mycorrhizal interactions is that different plants have different root surface flora (Table 6.1), and they may affect each other's rhizosphere including the mycorrhizas (Table 6.17). So the saprotrophic bacteria and fungi are interacting with each other and with the mycorrhizas (VAM in this case) and all are affected by neighbouring plants of the same or different species with their own microflora. Different plants have very different amounts of mycorrhiza; for example *Anthoxanthum* and

Table 6.16 Rhizoplane interactions between mycorrhizal fungi and bacteria. (From Bowen, G.D. & Theodorou, C., 1979, *Soil Biology and Biochemistry*, **11**, 119–26.)

	Length of root colonized by fungi (mm) **in the presence of bacteria**				
	Experiment 1			Experiment 2	
Mycorrhizal fungus	No bacteria	*Pseudomonas fluorescens*	*Bacillus* sp. 1	No bacteria	*Bacillus* sp. 2
Rhizopogon luteolus	27.6	8.3	23.7	20.7	34.1
Suillus luteus	16.2	3.2	8.7	19.1	26.2
Thelephora terrestris	22.4	12.1	18.1	12.1	11.2
Corticium bicolor	17.2	3.2	15.3	15.8	27.1
Pisolithus tinctorius	6.9	5.7	4.2	6.9	1.9
LSD $P<0.05$		5.4		6.2	
$P<0.01$		7.1		8.2	

Trifolium and *Lolium* versus *Trifolium* in Table 6.17. However *Trifolium* VAM are reduced when growing near *Anthoxanthum* but are not affected by *Lolium*. This interaction with the species of higher plant is also shown by saprotrophic bacteria and fungi which may be increased or decreased depending on whether the host plants are growing in single species stands or in mixtures.

VAM are important in phosphorus uptake, so a plant can affect the neighbour's phosphate status by interaction with its root microflora. VAM may also connect roots of different plants of the same or different species: hyphae have been traced through soil and shown to connect the roots. Phosphate is translocated from plant to plant, especially when one is nutrient rich and one is deficient, dead or senescent. This gives a closed cycling of phosphorus, without waiting for microbial decay of a dead neighbour to release

Table 6.17 The abundance of root surface bacteria and fungi and of mycorrhizal infection in glasshouse experiments. The two species were grown either in separate pots (alone) or together (mixed). Within any group of four figures (two species, alone and mixed), any two figures not followed by the same letter are significantly different ($P<0.05$) by Duncan's multiple-range test. (From Christie, P., Newman, E.I. and Campbell, R., 1978, *Soil Biology and Biochemistry*, **10**, 521–7.)

	Bacteria (cover %)		**Fungi** (mm mycelium mm^{-2} root surface)		**Mycorrhizas**: intensity of infection*	
	alone	mixed	alone	mixed	alone	mixed
Anthoxanthum	0.64b	0.99c	0.8a	2.7b	0.39a	0.49ab
Lolium	0.31a	0.28a	1.4a	1.4a	0.60b	0.54b
Anthoxanthum	0.64b	0.82c	0.8a	3.1b	0.39a	0.54a
Trifolium	0.48ab	0.41a	9.2c	3.4b	1.07c	0.90b
Lolium	0.31a	0.67b	1.4a	4.1b	0.60a	0.62a
Trifolium	0.48ab	0.24a	9.2c	4.2b	1.07b	1.09b

* Arbitrary units, scaled 0–4.

nutrients. The living plant can extract at least some of the available phosphorus for its own uses in a matter of a few days.

Some of the bacteria which are influenced by mycorrhizas may be free-living nitrogen-fixing and plant hormone producing organisms such as *Azotobacter* (p. 119 – 122). Inoculated *Azotobacter* may persist longer on mycorrhizal plants and exert a synergistic effect on plant growth. Double inoculated plants (*Glomus* and *Azotobacter*) have more saprotrophic bacteria in their rhizosphere, although fungi are unaffected or depressed, especially by *Azotobacter*.

Such experiments as these combined with other factors known to affect colonization of mycorrhizas, saprotrophs and pathogens make it amazing that anything so crude as simply inoculating a plant with a micro-organism can have an effect in field conditions. However, in some circumstances it is possible to get increased plant yields by such simple procedures.

There is a different sort of reaction between mycorrhizas and symbiotic nitrogen-fixing organisms which form root nodules, for both the root nodule organisms and the mycorrhizas are a drain on the photosynthate of the plant and could be in competition. Again the combined symbiosis may be synergistic, the plant growing better with both than the sum of the separate effects, possibly because the nitrogen-fixation has a high phosphorus demand, so the two are closely linked. Figure 6.18 shows the distribution of fixed carbon in shoots and roots as a proportion of total photosynthate. Though the proportion of carbon going to root and shoot remains more or less constant regardless of the symbionts the fixation rate increases so that the carbohydrate drain of the micro-organisms is to some extent compensated for, though the

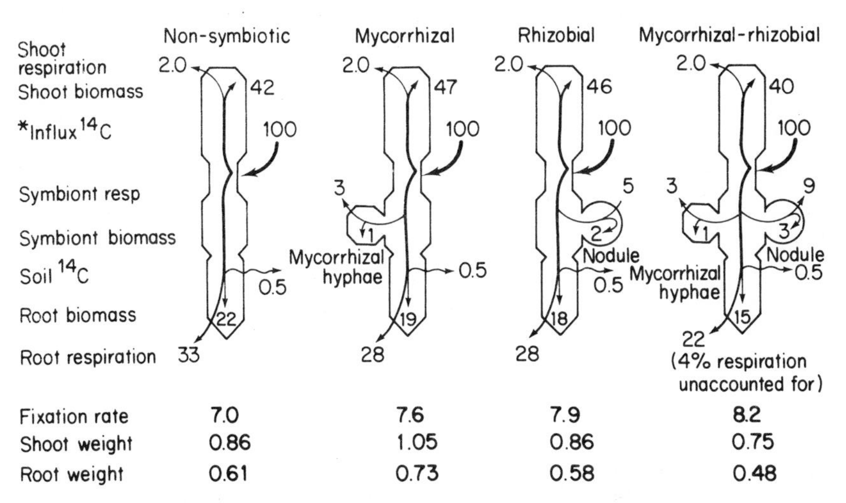

	Non-symbiotic	Mycorrhizal	Rhizobial	Mycorrhizal-rhizobial
Fixation rate	7.0	7.6	7.9	8.2
Shoot weight	0.86	1.05	0.86	0.75
Root weight	0.61	0.73	0.58	0.48

Fig. 6.18 Distribution of photosynthate in nodulated, mycorrhizal roots of faba beans (*Vicia faba*). The ^{14}C flow to various compartments of symbiotic faba beans (4–5 weeks old) after shoots were exposed above ground to $^{14}CO_2$ under continuous light. The fixation rate is expressed as mg of carbon per gram of shoot per hour. The shoot weight and root weight are expressed as g of carbon. The carbon influx has been equalized to 100 units of carbon per g of shoot carbon. (From Paul, E.A. & Kucey, R.M.N., 1981, *Science,* **213**, 473–4. Copyright 1981 by the American Association for the Advancement of Science.)

overall plant weight is reduced (though not significantly) in the presence of the two symbionts compared with fertilized, un-infected plants. Notice that the nodule respiration increases in the presence of mycorrhizas and nitrogen-fixation was also increased, as noted above.

The occurrence of ectomycorrhizas may protect against root pathogens such as *Phytophthora cinnamomi* in pine, and VAM may also do this if established on the roots prior to the attack by *P. cinnamomi*. The method of protection may be a mechanical barrier of the sheath, competition for nutrients, or by affecting the rhizosphere population or by the production of toxins. More than 100 ectomycorrhizas are known to be able to produce antibiotics, including *Leucopaxillus* which reduces infection of *P. cinnamomi*. There are many diseases which it is claimed are reduced in the presence of mycorrhizal infection but whether this is one of the direct effects above or whether it is via an effect on general plant vigour is not clear. This latter point is important in foliar diseases which, in contrast, may be more serious in plants infected with VAM; *Erysiphe graminis* and virus diseases may both be more serious, probably because improved phosphorus nutrition makes the plant a better food source.

Interactions amongst saprotrophs and pathogens

The main interactions between micro-organisms have been discussed previously (p. 14). We are concerned here with predation, parasitism and various forms of antagonism such as competition and antibiotic production. These interactions have been shown to be important in soil in general and in particular on or near roots. They may result in reduced numbers or activities of pathogens as we have seen with mycorrhizas. Such control of disease is a type of biological control which also includes any means of disease reduction which ultimately depends upon biological means; for example cultivation practices favouring antagonistic saprotrophs are considered to be biological control.

Bacteria are subject to predation by other bacteria (*Bdellovibrio*) and by some protozoa. These interactions are difficult to study in any natural system because of the spatial heterogeneity and the complexity. Even if constant bacterial numbers are recorded it is never certain whether this is a static population or whether it is a balance of predation against reproduction. There are therefore only data from simplified microcosms with one or a few species of bacteria and protozoa. These studies suggest that bacterial numbers are frequently suppressed by protozoa though the mineralization of nitrogen and phosphorus and the oxygen consumed are increased as nutrients mobilized by the bacteria are released. Though these studies were sometimes designed to mimic the rhizosphere there is no direct evidence that they actually occur in the rhizosphere or on roots in natural soil, though it seems plausible that they do. Myxamoebae also occur in normal rhizosphere (Table 6.6) and are predatory on bacteria.

Amoebae of various sorts (often identified as vampyrellid amoebae) have also been shown to attack fungal hyphae and spores in soil and on plant roots. The amoeba contacts the fungal structure and in some way erodes an annular depression in the wall, eventually penetrating right through, cutting out a disc of wall, which is then discarded, and leaves a circular hole (Fig. 6.19A). The amoeba enters the cell and consumes the contents. Many fungi are attacked, including root pathogens such as *Gaeumannomyces graminis* (p. 143) and amoebae have been recovered from soils which are suppressive, i.e. do not allow development of the disease. The circumstantial evidence is good, but again there is no positive proof that such organisms significantly control disease under natural conditions.

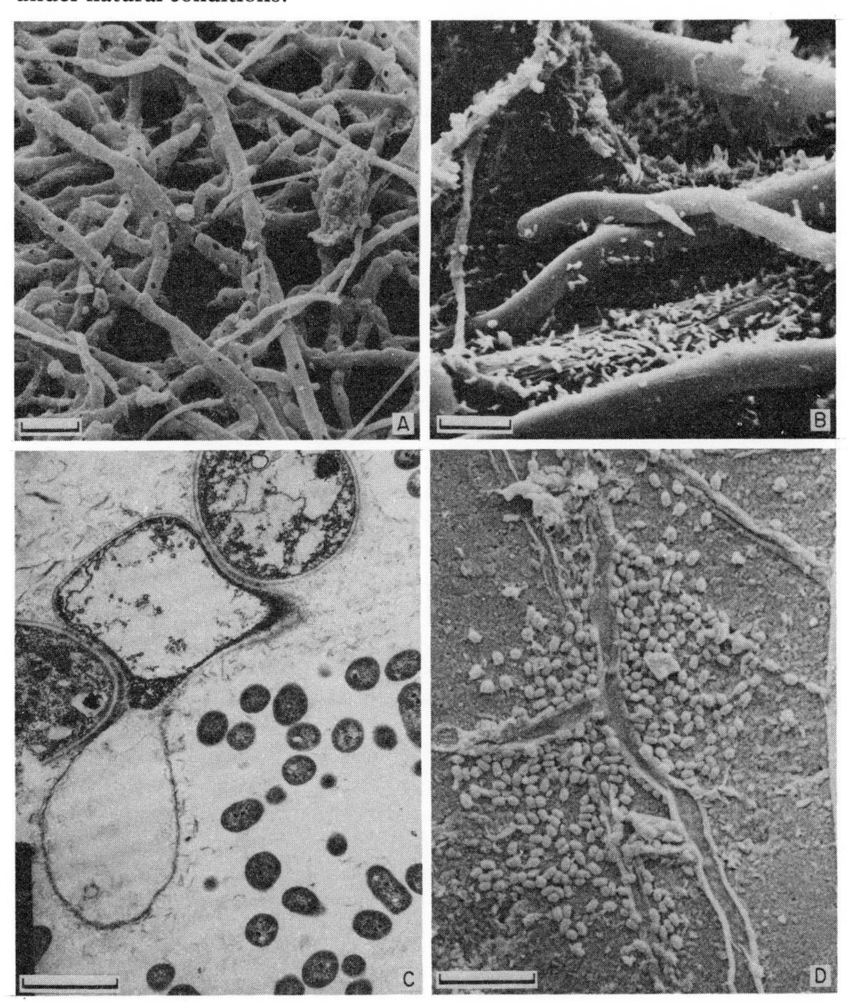

Fig. 6.19 Interactions between micro-organisms on roots. **A.** Scanning electron micrograph of perforations in hyphae caused by vampyrellid amoebae following burial in the soil for 8 weeks. Bar = 20 μm. (From Homma, Y., Sitton, J.W., Cook, R.J. & Old, K.M., 1979, *Phytopathology*, **69**, 1118–22.) **B.** Hyphae with bacteria attached end-on. There are also many bacteria on the root surface. Bar = 5 μm. **C.** Bacteria associated with lysed hyphae on the root. There is no contact between the bacteria and the fungus. Bar = 2 μm. **D.** A hypha lysed while growing on the gas-liquid interface of a water film in the soil. Bar = 10 μm. (From Campbell, R., 1983 *Canadian Journal of Microbiology*, **29**, 39–45.)

Hyphae and bacteria interact, by causing lysis for example. Bacteria become attached to the hyphae, often end-on (Fig. 6.19B) and the hyphal contents become vacuolated and finally disappear (Fig. 6.19C). There are many examples of such interactions but some bacteria are known to do this to *G. graminis* and they give some field control of the disease. *P. cinnamomi* may also be lysed by bacteria. As a variation, *Rhizobium* inoculated for nodulation and nitrogen fixation can protect legumes from *Phytophthora megasperma* and other fungi, but this has only been shown in culture, not under field conditions. Bacteria can also lyse hyphae without actual contact, presumably by the production of water soluble toxins or enzymes (Fig. 6.19C and D). The reverse process occurs in the production of edible mushrooms (*Agaricus bisporus*, p. 172) which use bacterial cells, produced during composting, as a food source. *A. bisporus* can live on bacteria as a sole carbon and nitrogen source.

Finally in these possible interactions we have to consider those between hyphae of different species. This may be again at a distance so that one hypha releases a toxin which lyses the other species ahead of it as in the case of *Peniophora* (= *Phlebia*) antagonizing *Heterobasidion* (p.144): this has been called hyphal interference. Alternatively hyphae may coil round or penetrate other hyphae which they are parasitizing. This is the case when *Trichoderma harzianum* is used as a control agent for *Rhizoctonia* attacking roots. In either case the 'host' hypha is killed.

There may also be competition in the rhizosphere for nutrients and for available space (i.e. space where environmental conditions are suitable for growth) between all sorts of micro-organisms. This has led to the use of non-virulent organisms, closely related to pathogens, to fill the same or very similar niche so eliminating the pathogen by competitive exclusion.

We have therefore a very wide range of interactions which are possible and have been reported for some pathogen or other, often under laboratory conditions. The question is, can this be put to use on a field scale to help control root pathogens that are difficult to treat with fungicides? Some do work in the field; the *Peniophora/Heterobasidion* system is available commercially and others have been used in field trials. It is however very difficult to introduce an alien organism into the soil and have it persist. The only way that this can usually be achieved is either to use massive inoculum or to disturb the system in some way, by cultural practices or the application of partial sterilizing agents, before introducing the antagonist into the system. This combined use of chemical and biological means of control is known as integrated control. Biological control of root diseases may also be possible in horticulture, which is an intensive form of plant cultivation, where environmental conditions can often be controlled. Sterile or almost sterile composts are also used and the alien antagonist can be introduced into this with a reasonable chance of survival. This is similar to introducing *Peniophora* into freshly cut stumps or wounds where it will become established in the absence of competition.

Biological control (Cook & Baker, 1983) of root pathogens is an extensive subject with many practical difficulties but it is nevertheless making some headway. Its most likely success will be in specialist situations and as a part of integrated control, especially with the direct introduction of antagonists.

Many agricultural practices which work by biological means will continue to be extensively used: these include rotations, cultivation systems, green manuring, etc.

Conclusion

It should now be obvious that the microbiology of root surfaces is very complex, even though we really are only at the beginning of the studies. There is now a working knowledge of the microbes present and the processes occurring in temperate crop plants, but the generalities and concepts developed for these need to be checked and extended to tropical crops and all natural vegetation. We know very little about microbial interactions on roots under anything approaching natural conditions. There are hints that such interactions may be important in controlling economically-important root pathogens and could be used agriculturally if only we knew enough to influence and manipulate the root microflora in any but the crudest ways which are used now. The whole picture of root pathology will change if or when downward (phloem) translocated fungicides are available commercially. The manipulation of the microflora to encourage advantageous symbionts will also develop enormously as artificial fertilizers become more expensive. The present approach of adding inoculum works in many cases, but more could be done in strain selection, possibly by genetic engineering and in ensuring an understanding of how added inoculum competes and survives in the soil. These mutualistic symbionts do however use host photosynthate so it must be kept in mind that the price paid in yield for the nitrogen or phosphorus, should not exceed its value to the plant.

Root microbiology in realistic soil conditions will continue to be an expanding, if technically difficult area of research.

Selected references and further reading

Campbell, R. (1983). *Microbial ecology*, 2nd edition. Blackwell Scientific Publications, Oxford. pp. 191.

Cook, R.J. & Baker, K.S. (1983). *The nature and practice of biological control of plant pathogens*. American Phytopathological Society, St. Paul, Minnesota. pp. 539.

Foster, R.C., Rovira, A.D. & Cock, P.W. (1983). *Ultrastructure of the root-soil interface*. American Phytopathological Society, St. Paul, Minnesota. pp. 168.

Garrett, S.D. (1970). *Pathogenic root-infecting fungi*. Cambridge University Press, Cambridge. pp. 294.

Harley, J.L. & Scott Russell, R. (Eds) (1979). *The soil-root interface*. Academic Press, London. pp. 448.

Harley, J.L. & Smith, S.E. (1983). *Mycorrhizal symbiosis*. Academic Press, London, pp. 483.

Lynch, J.M. (1983). *Soil biotechnology: microbiological factors in crop productivity*. Blackwell Scientific Publications, Oxford. pp. 191.

Mosse, B., Stribley, D.P. & Lepacon, F. (1981). Ecology of mycorrhizae and mycorrhizal fungi. *Advances in Microbial Ecology*, 5, 137–210.

Newman, E.I. (1978). Root microorganisms: their significance in the ecosystem. *Biological Revue*, **53**, 511–44.

Postgate, J.R. (1982). *The fundamentals of nitrogen fixation*. Cambridge University Press, Cambridge. pp. 252.
Schippers, B. & Gams, W. (Eds) (1979). *Soil-borne plant pathogens*. Academic Press, London. pp. 683.
Sprent, J.I. (1979). *The biology of nitrogen-fixing organisms*. McGraw-Hill, London. pp. 196.
Suslow, P.V. (1982). Role of root-colonising bacteria in plant growth. In Mount, M.S. & Lacy, G.H. (Eds). *Phytopathogenic prokaryotes*. Academic Press, New York. pp. 541.
Waid, J.S. (1974). Decomposition of roots. In Dickinson, C.H. & Pugh, G.J.F. (Eds). *Biology of plant litter decomposition*, Vol. 1. Academic Press, London. p. 175–211.
Woldendorp, J.W. (1978). The rhizosphere as part of the plant-soil system. In *Structure and functioning of plant population*. Verhandeligen der Koninklijke, Nederlandse Akademie van Wetenschappen, Afdeling Natuurkunde, Twede Reeks, deel 70.

7
Decomposition of Plant Litter

Introduction

We have already considered some aspects of decomposition when discussing senescence of leaves and roots (p. 58 and 118), the microbiology of wood (p. 95) and various post-harvest decays of agricultural products. This chapter will consider (1) some general principles of litter decay, including interactions with other organisms such as arthropods, (2) agricultural and industrial litter decomposition systems such as silage production and (3) the microbiology of the rumen, caecum and the guts of termites which are the basis of grazing food chains and herbivory in general.

The cycling of nutrients, which is largely dependent on microbial activity, is considered in many texts (Fenchel & Blackburn, 1979; Campbell, 1983). The decomposition of plant litter (Dickinson & Pugh, 1974) in both aquatic and terrestrial systems (Anderson & Macfadyan, 1976; Swift, Heal & Anderson, 1979) is also extensively studied and many data on standing crops, productivity, turnover times, etc., have been produced as a result of the detailed studies of different biomes in the International Biological Programme.

Organic matter and trophic structure in terrestrial ecosystems

In most of these natural habitats the litter is decomposed *in situ* so that the quantity is limited by the productivity of the site. In agriculture much of the material is removed as crop and subsequently deposited elsewhere, often in a concentrated form, where it may cause problems with excessive biological oxygen demand necessary for its decay. In these agricultural systems the productivity is greatly influenced by the amount of carbon, nitrogen and minerals added back to the soil as manures and fertilizers. Agricultural soils generally have low available organic matter and most microbes in such soils are carbon limited.

Comparing productivity of different habitats is very difficult because of the many assumptions inherent in the estimates and the different growth forms, etc., of the plants (see Swift *et al.*, 1979, for a more complete listing). However on an annual basis the least productive habitats are dry deserts, then polar tundra of various sorts (net primary production is 1–5 t ha^{-1} y^{-1}), though for short spells during the growing season they may be very productive. Temperate forests are next, generally producing 4–20 t ha^{-1} y^{-1}, though occasionally more, and subtropical and tropical rain forests may go up to 30 t ha^{-1} y^{-1}. Net primary production of crop plants is very variable but in general it is in the upper range of those just mentioned, provided that fertilizer applications are maintained. There are a few exceptional crops such as sugar-cane which have a very high productivity.

Obviously the different vegetation types not only produce different quantities of carbon but different sorts of plants. Production may be of leaves only or leaves and twigs or large woody stems to mention only the above ground portions: the resource quality varies in different habitats. The dead plant material may become immediately available for decomposition or it may be stored in the plants themselves. Most grasslands have a very small standing biomass but forest devotes 20–60% of its net primary production to maintaining and producing its standing biomass. Terrestrial systems are grazed by animals but usually only to a slight extent and even in grassland only about one quarter of the biomass is removed by herbivores. The remainder is decomposed by micro-organisms and even that eaten by animals is decomposed by microbes before it is used by the animal itself (p. 165 – 70). In fact the breakdown of the major plant constituents is all done microbially. No animal, except perhaps some snails, has been shown to produce cellulase. The bodies of the herbivores and carnivores which feed on them are also decomposed by micro-organisms.

It should be stressed that plant litter also includes roots which die and are renewed (p. 118). A considerable proportion of the litter is formed below ground, the amount again varies with vegetation type, with tundra and grassland the highest (70–85%) and various forest types only contributing about 20% of their litter below ground.

Normally most of the terrestrial plant remains, which are of course already colonized by micro-organisms, form a litter layer on or in the soil and are there decomposed. The process is quite rapid at first but as more and more intractable substances are left the decomposition slows and finally a brown, amorphous substance, humus, is produced, partly from the original plants and partly from the microbes. This may take hundreds or even thousands of years to be broken down (Fig. 7.6).

The soil organic matter is therefore dependent not only on the vegetation type but on the rate and type of decomposition processes in the soil which may be determined by climate, drainage patterns, soil pH, the soil minerals and many other factors. These can interact to produce an almost infinite variety of types of organic matter in the soil, but two extremes are usually recognized with many intermediate conditions. Mor organic matter is characterized by having recognizable plant parts, a low pH (3.5 – 5.0), a C:N ratio of more than 20 and a soil of low base saturation. It is common on well-drained sites in regions of high rainfall where leaching is pronounced and the vegetation is either ericaceous or coniferous, producing an acid litter. Mull organic matter in contrast is usually amorphous with a lower C:N ratio and a higher pH and on a soil with greater base saturation. It is characteristic of broadleaved woodland and grassland plant communities.

These differences are of course reflected in the microbiology: mor is dominated by fungi and small arthropods (the mesofauna) which usually live on the micro-organisms. Mull has higher bacterial counts and many earthworms (macrofauna), the fauna eats litter and breaks it up into small pieces before the main microbial attack. The kinds of animals and micro-organisms are determined by such factors as the resource quality, including the available nutrients and the presence of toxins. Animals and micro-organisms may show resource specificity, only using some components of the resource. Such organisms determine which resources decompose, how they decompose and which other organisms do the decomposition.

Soil animals and micro-organisms are a very diverse group and have for convenience been divided into categories based on size. Slightly different results are obtained depending on the dimension chosen but Fig. 7.1 illustrates one scheme with the micro-flora and micro-fauna, including bacteria, fungi, protozoa, nematodes and rotifers. The mesofauna includes mites and collembola and the macro- and mega-fauna such animals as woodlice, beetles, molluscs, millipedes, etc. Since the distribution of the different microbes and animals varies with organic matter type it follows that they will vary in importance with vegetation type and latitude, for mor humus is most common at high latitudes (Fig. 7.2). This is obviously a great simplification and much study has been devoted to the distribution of particular groups of microbes or fauna: for example ciliates and testate amoebae have been studied (Fig. 7.3),

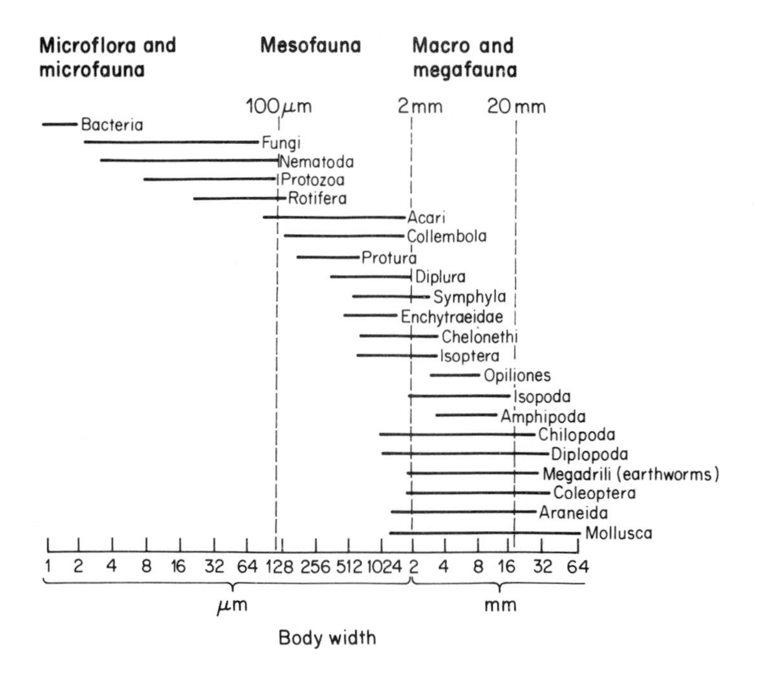

Fig. 7.1 Size classification of organisms in a decomposer food web, determined by body width. (From Swift, Heal & Anderson, 1979.)

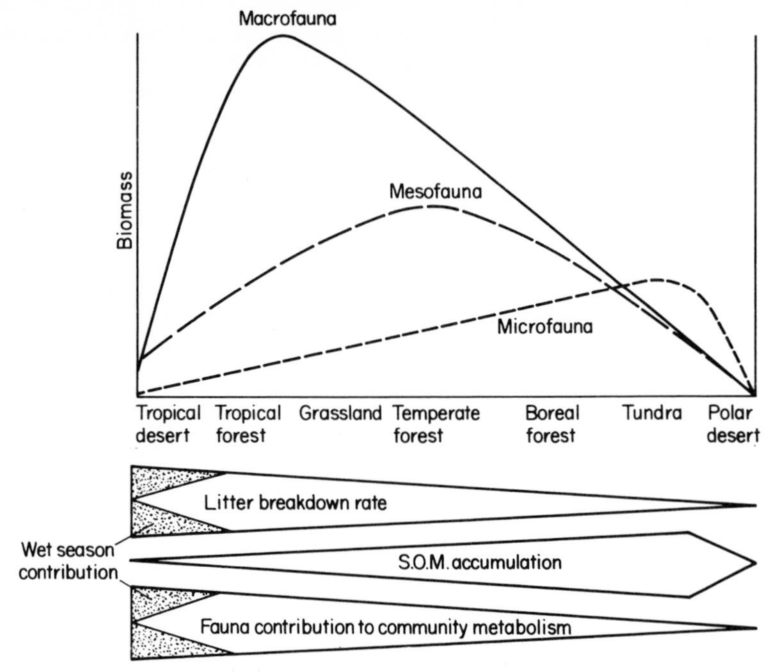

Fig. 7.2 Hypothetical patterns of latitudinal variation in the contribution of macro-, meso- and micro-fauna to total soil faunal biomass. The effect on litter breakdown rates of changes in the relative importance of the three fauna size groups is represented as a gradient together with the fauna contribution to soil community metabolism. The favourability of the soil environment for microbial decomposition is represented by the cline of soil organic matter (SOM) accumulation from the poles to the equator; SOM accumulation is promoted by low temperatures and water-logging where microbial activity is impeded. (From Swift, Heal & Anderson, 1979.)

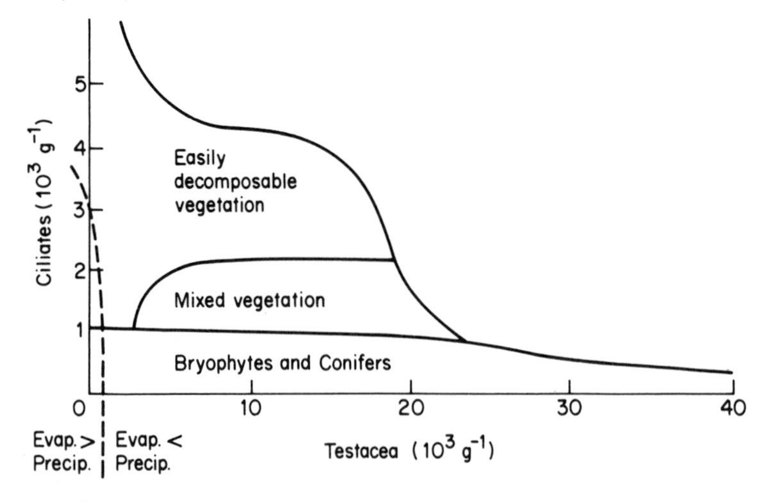

Fig. 7.3 The relationship of the number of ciliates and testacea in litters to the type of vegetation. The region to the left of the broken line denotes arid habitats. (From Bamforth, S.F., 1973, *American Zoologist*, **13**, 171–6.)

the latter are most common under acid bryophytes and conifers but both groups occur in more easily decomposable litter such as deciduous forest or grassland. Ciliates are bacterivorous and occur in those habitats with many bacteria, but the slower growing testacea depend on organic matter particles.

Detailed quantitative data are available for Scots pine forest (mor humus, Table 7.1). The fungi are dominant in biomass, but not in respiratory activity, and together with the bacteria they far outweigh all the other soil flora and fauna put together. As expected the rhizopods, which include testate amoebae, are the dominant protozoa and the micro-arthropods are more important than the macro-arthropods. These figures are annual means, the individual numbers vary throughout the year, usually in response to climate and events such as litter fall in deciduous woodlands. These data are unique and very valuable. They are the result of an enormous amount of work to study all groups of organisms in a single vegetation type over several years and to relate

Table 7.1 Number of identified species and annual means of hyphal length, abundance, biomass and respiratory metabolism of the organisms found in the 120-year-old Scots pine stand at Ivantjarnsheden according to the regular soil samplings. The annual means are calculated for the period July 1974–June 1975 (FDA-active fungi 1975–77 and protozoa from 1978 included). n.d. = not determined. (From Persson *et al.*, 1980.)

	No. of species	**Abundance** (no. m^{-2})	**Biomass** (mg dw m^{-2})	**Respiratory metabolism** (mg C m^{-2} yr^{-1})
Fungi (agar-film)[a]	51[b]	–[c]	80 000	60 000
Fungi (FDA)		–	1 400	
Bacteria[a]	n.d.	16×10^{13}	20 000	105 000
Protozoa[a]	n.d.	130×10^{6}	90	1 700
Ciliata	n.d.	3×10^{6}	5	69
Flagellata	n.d.	40×10^{6}	16	330
Rhizopoda	n.d.	86×10^{6}	69	1 300
Metazoan microfauna[a]	34	4 950 000	166	930
Nematoda	28	4 390 000	122	740
Rotatoria	2	510 000	33	140
Tardigrada	4	49 000	11	43
Enchytraeidae	3	16 200	405	2 000
Microarthropods	118	744 000	609	1 800
Symphyla	1	35	0.2	1
Collembola	26	60 000	94	470
Protura	1	700	0.3	3
Acari	90	684 000	515	1 340
Gamasina	14	8 400	47	130
Uropodina	2	180	1.3	2
Prostigmata	20	210 000	50	210
Astigmata	2	40 000	29	110
Cryptostigmata	52	425 000	388	890
Macroarthropods	>61	1 700	313	720

[a] Annual means underestimated here because of insufficient sampling depths.
[b] Microfungal species (Söderström, B.E. & Baath, E. 1978. *Holarctic Ecology*, **1**, 62–72).
[c] Average hyphal length 85 000 km m^{-2} (before correction to greater depth).

these data to soil characteristics and chemistry. They only tell us about one ecosystem and it is unsafe to generalize, but they point out what ought to be done elsewhere.

These size and taxonomic criteria are one way of considering decomposition but the matter can also be looked at in terms of trophic structure and nutrition which cuts across these other categories. There are biotrophs living on the live plant and we have considered these in other chapters. Necrotrophs include plant parasites and also animals feeding off and killing fungi, bacteria and protozoa whose population density and structure they may influence (see p. 113). Obviously the main groups we are concerned with in litter decomposition are the saprotrophs, those organisms living on dead material. It may be the dead plant, or the dead micro-, meso- or macro-fauna and -flora. Bacteria and fungi are the main primary decomposers which are eaten by a large range of micro- and meso-fauna some of which actually encourage the growth of microbes on litter. The biomass and respiration of these different groups of soil micro-fauna varies down the profile, and on an annual basis reflecting the variation of their food source. On an annual basis (Fig. 7.4) the bacterivores and fungivores are most important in terms of respiration and biomass respectively. These data can then be built up into carbon flow charts for plant litter layer in this forest (Fig. 7.5). Notice again the importance of bacterial and fungal biomass, and the carbon flow which derives from it in the form of dead microbial organic matter and that eaten by animals. This emphasizes the small proportion of the litter which is taken directly by saprovores and herbivores. The faeces of such animals also form an important part of the litter or may be a major component of some soils and they are actively colonized by the micro-organisms. There is an improvement in the

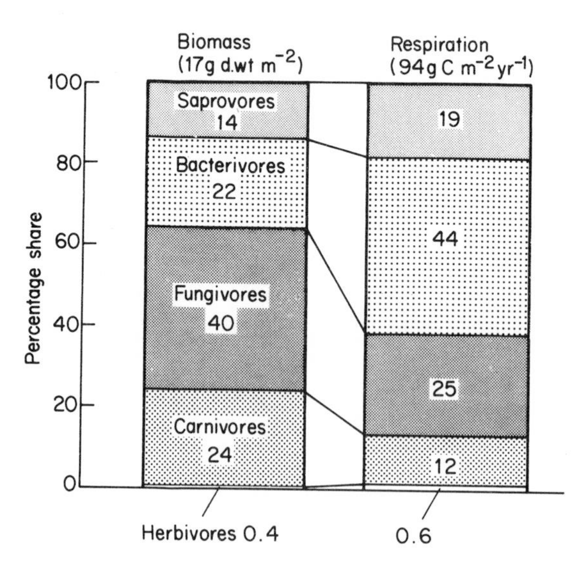

Fig. 7.4 Contribution (%) to the annual mean biomass and annual respiration of trophic groups within the soil fauna. Note that the bacterivore share is very approximate. (From Persson, *et al.*, 1980.)

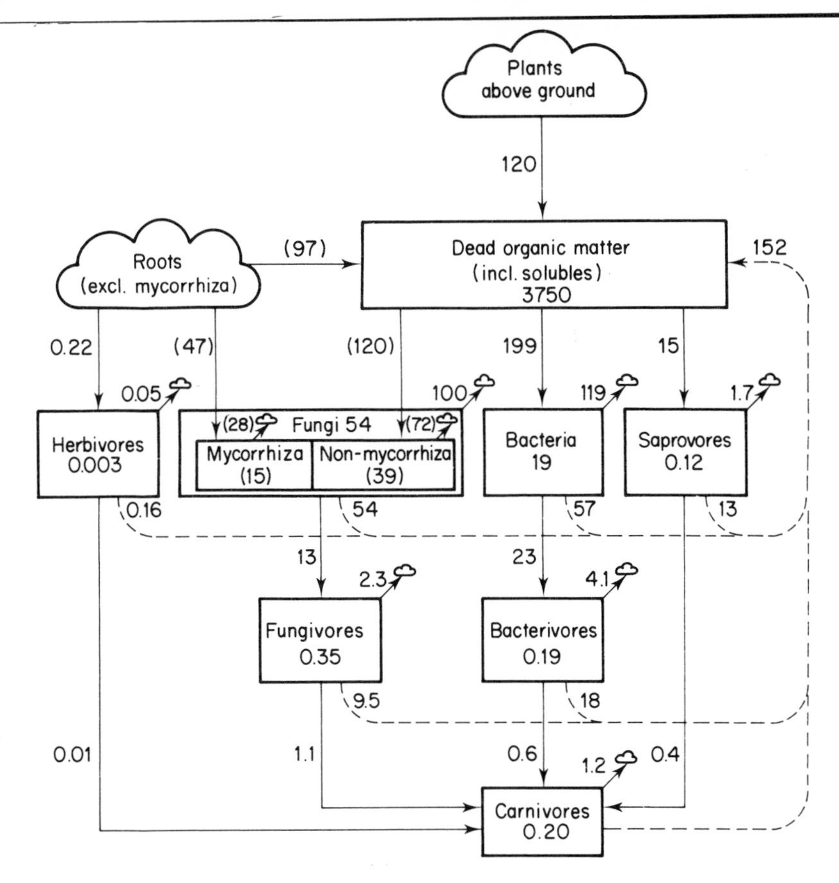

Fig. 7.5 Carbon flow diagram of the soil in the 120-year-old Scots pine stand at Ivantjarnsheden to a depth of 30 cm in the mineral soil. Boxes denote annual mean biomass (g C m^{-2}) and arrows denote flows (g C m^{-2} y^{-1}). Arrows with small clouds indicate respiration. Broken lines ending in an arrow denote transfer of dead, non-predated organisms and faeces to the pool of dead organic matter. All values are uncertain and in particular those influenced by the assumptions of mycorrhiza (within parentheses). (From Persson, *et al.*, 1980.)

resource quality from plant litter to primary saprotrophic meso- and macrofauna. The animals are incapable of living without microbes, whether they have them outside their bodies on the litter or within their gut, and many of the animals are coprophagous, using the microbial breakdown of their own faeces as a food source. Faeces and the organisms which produce them may therefore be a very important part of the decomposition system, greatly increasing the rate of mineralization. Microbes alone can decompose litter but usually slowly.

All the above information refers to natural or semi-natural ecosystems. If we consider arable agricultural systems there is generally no well-developed soil profile, for the ground is disturbed by ploughing or other cultivation giving a fairly uniform distribution of organisms and organic carbon in the upper soil layers. However recent developments in western agriculture, especially in the intensive growing of cereals, and the increasing costs of fuel and labour have led to the development of minimum tillage systems. Organic matter and the associated nutrients accumulate on the surface and in the upper horizons. If quantities are large enough its decomposition can create anaerobic conditions which inhibit seed germination and may be toxic to root growth. Alternatively, or in addition, the cycling of nitrogen and phosphorus may occur above the root zone so that nitrogen applied as a fertilizer is immobilized and more must be put on to reach the growing crop. Some of these effects are illustrated in Table 7.2: notice the more even distribution in relation to depth

Table 7.2 The numbers of bacteria and fungi and the rates of respiration at two depths in a clay soil sown to spring barley, either by direct drilling or after ploughing. (Modified from Agricultural Research Council, Annual Report *Letcombe Research Station*, Wantage, England, 1976.)

	Ploughed		**Direct drilled**		**LSD**
Depth (cm)	0–5	10–23	0–5	10–23	$P<0.01$
Number bacteria (10^6 g^{-1})	1.71	1.52	1.82	0.92	0.98
Number fungi (10^4 g^{-1})	2.35	1.45	3.47	0.73	0.71
Oxygen uptake (μl h^{-1} g^{-1})	9.9	7.1	9.0	1.0	5.7
Carbon dioxide (μl h^{-1} g^{-1})	1.9	1.5	2.8	1.6	0.1
Organic matter (%)	3.9	3.0	5.2	2.2	1.5
Extractable PO_4^{2-} (μg P g^{-1})	36.8	23.8	57.4	3.6	—
Extractable K (μg K g^{-1})	318	350	536	250	—

of all the parameters in the disturbed, ploughed soil and the concentration of nutrients, organisms, oxygen demand and organic matter in the surface layer of the direct drilled plot. In most cases there is not significantly more or less of anything, merely a different distribution of the same material. The unploughed land resembles more closely natural conditions with a distinct organic matter horizon. The latter may be detrimental to crops, not only because of the immobilization of nutrients, but also it may encourage the survival of trash-borne plant pathogens. Because of these and other problems intensive cereal growing in Britain is now tending to plough or otherwise disturb the soil every few years, especially on heavy soils. So direct drilling is used until the detrimental effects become uneconomic, then ploughing is done for one year before another period of no-plough. Some of these problems can be alleviated by removing the straw (in the case of cereals) by baling and use as animal bedding, or more usually by burning. Baling is often not economic if there is no obvious use for the straw on a specialized cereal farm, as opposed to the traditional mixed farm with livestock. In dry-land wheat cultivation in western U.S.A. minimum tillage may be used to reduce erosion and retain soil moisture.

It should be stressed that direct drilling changes the distribution of material, not usually the total quantity. Indeed it is very difficult to change soil organic matter very much by agricultural practices. Table 7.3 shows results of the incorporation of different forms of organic matter on the total soil organic matter content. It is difficult to change the amount of organic matter over a twenty-year period by more than 0.1 or 0.2 per cent with the normal amounts of straw and farmyard manure available. This is not to say that in the long term (hundreds or thousands of years) there may not be an effect, only that nothing dramatic is likely to happen in the lifetime of one farmer. This is partly because of the relatively small quantities of organic matter that are added to the soil surface and incorporated. Furthermore what is added is respired rapidly by the carbon-limited soil organisms. The main factor however is the long turnover time of organic matter in soil which, for the more resistant fractions, may be more than 1000 years (Fig. 7.6) The fresh organic matter is rapidly attacked but the small proportion which is a resistant residue takes very much longer. Even though only a small proportion of the new organic matter reaches the fulvic acid and humus fractions these constitute the majority of the soil carbon because of slow build-up over the life of the soil. It takes many years to make any significant change in these major soil organic matter pools, whether this be a slow loss by stopping addition or an increase by the use of large quantities of organic manures.

Micro-organisms involved in decomposition

Let us now consider some of the more important organisms involved in litter decay (Fig. 7.1) in more detail, concentrating on the micro-organisms and their interactions with the animals. Remember that we are not talking about this or that organism decaying litter: complete breakdown is dependent on many interacting organisms.

Table 7.3 Effects of ploughing-in straw or farmyard manure on soil organic carbon. (From Russell, E.W., 1977, *Philosophical Transactions of the Royal Society, London*. B. **281**, 209–19.)

	Percentage of carbon in soil after 18 years of a 6-course rotation at experimental husbandry farms					
	Initial value	Farmyard manure	Straw ploughed in	Straw removed or burnt	Effect of farmyard manure	Effect of straw
Boxworth (calcareous clay)	1.67	1.68	1.61	1.51	0.17	0.10
Gleadthorpe (loamy sand)	1.20	1.24	1.21	1.17	0.07	0.04
High Mowthorpe (chalky silty loam)	2.26	2.24	2.18	2.09	0.15	0.09
Terrington (silty loam)	1.47	1.38	1.35	1.35	0.03	0.00

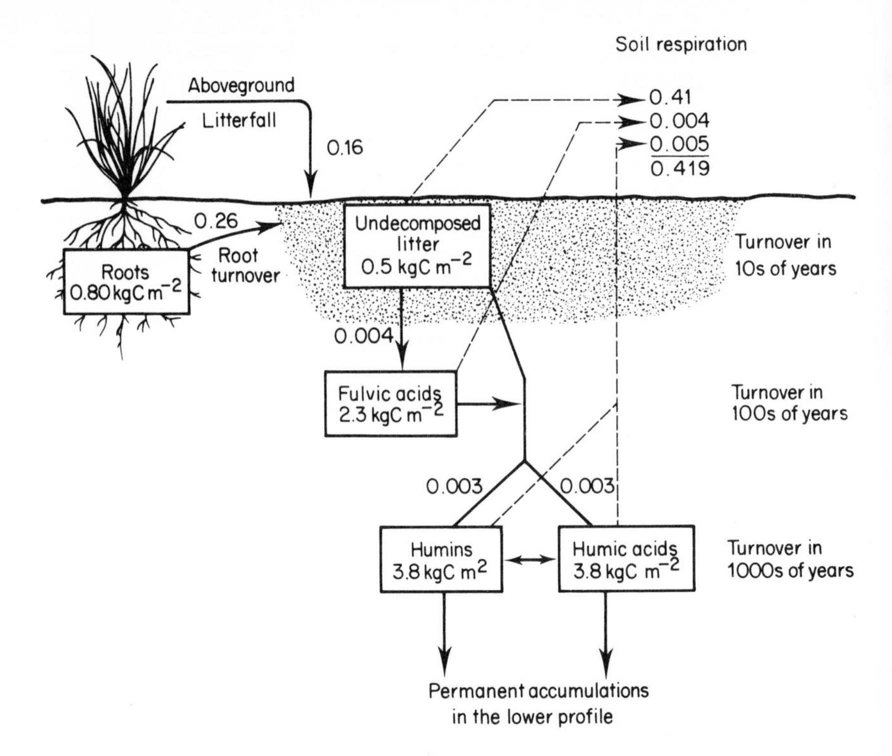

Fig. 7.6 Detrital carbon dynamics for the 0–20 cm layer of a Chernozem grassland soil. Carbon pools (kg C m^{-2}) and annual transfers (kg C m^{-2} y^{-1}) are indicated. Total profile content to 20 cm is 10.4 kg C m^{-2}. (From Schlesinger, W.H., 1977. Reproduced, with permission, from the *Annual Review Ecology and Systematics*, **8**, 51–81. © 1977 Annual Reviews Inc.)

Bacteria and fungi

These are the primary sources of the enzymes necessary to degrade the litter. They are basically different in growth form: fungi can penetrate gross structures whereas bacteria can only colonize surfaces, though their small size allows them to enter through even very small holes. Fungi are therefore more important when the litter is in large pieces, especially as they are usually tolerant to lower pH, in acid peat for example, and to dryer conditions, as in the surface leaf litter. Bacteria are particularly important under anaerobic conditions, with finely-divided organic matter and with more recalcitrant substrates: the small sized particles having been comminuted and 'processed' by others may well have had most of the easily available nutrients removed. Many of the resources reaching the litter are already colonized by *r* strategists, and the primary sugar fungi (p. 92) have probably already been replaced by cellulose and lignin decomposers. The associated secondary sugar fungi (p.92) and bacteria live off the sugars, etc., produced by the extracellular hydrolysis done by the basidiomycetes and the ascomycetes. Bacteria are also accociated with the outside of the hyphae of all sorts of fungi.

There is stratification within the soil profile in the general biomass and trophic and nutritional groups of organisms. In mor soils there are usually three litter horizons defined, the layer of intact leaves (called the L or A_{00}), the layer of broken but still recognizable pieces (F, A_{01} or O_H) and the amorphous organic matter (H, A_{02} or A_H). Basidiomycetes and some ascomycetes are characteristic of the layer of small fragments where they often form a dense weft of mycelium, mycelial fans or strands. The amorphous layer tends to have more bacteria. Particular genera of fungi are more common in some horizons than others (Fig. 7.7). There is a distinct zonation with heavily sporing types near the surface

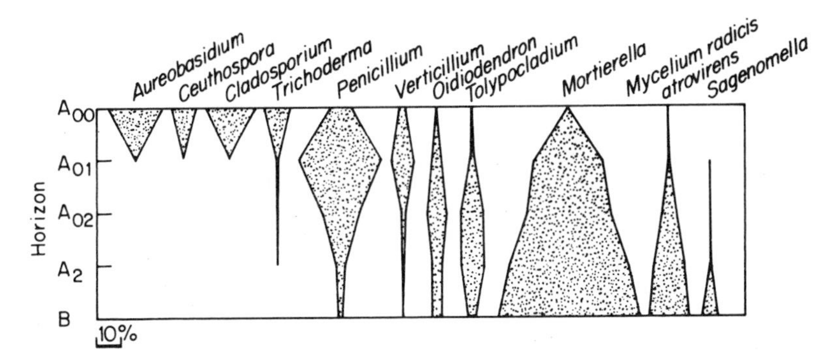

Fig. 7.7 Distribution of some micro-fungal genera in different soil horizons in the Scots-pine stand at Ivantjarnsheden (% of total number of isolates per horizon). (From Persson *et al.*, 1980.)

where the desiccation resistant structures are important, e.g. the chlamydospores of *Aureobasidium*. Lower down in the amorphous humus specialized fungi able to degrade recalcitrant molecules occur; for example *Mortierella* and non-sporing types become more important. Even within the single genus *Mortierella*, the different species may show zonation in the litter layers (Fig. 7.8).

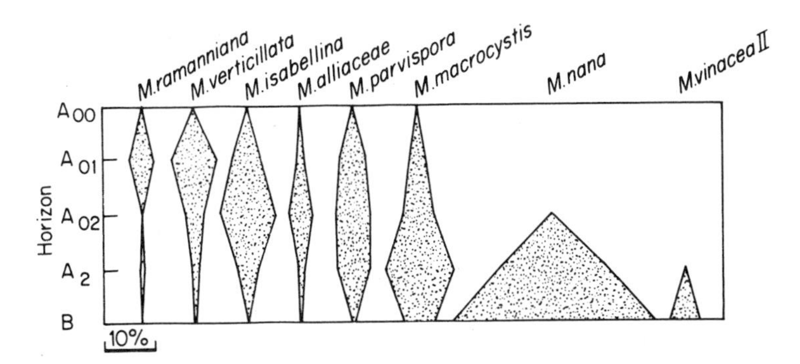

Fig. 7.8 Distribution of some species of *Mortierella* in different soil horizons in the Scots pine stand at Ivantjarnsheden (% of total number of isolates per horizon). (From Persson *et al.*, 1980.)

Table 7.4 Comparison of the distribution of the biomass of basidiomycetes (kg ha^{-1} dry weight) with that of other microbial decomposers in the floor of a temperate deciduous woodland with mull humus (Meathop Wood, Cumbria, U.K.) (From Frankland, J.C., 1982, in *Decomposer Basidiomycetes: their biology and ecology*. Frankland, J.C., Hedger, J.M. & Swift, M.J. (Eds). Cambridge University Press, Cambridge. 241–61.)

<table>
<tr><th rowspan="2">Substrate or horizon</th><th rowspan="2"></th><th colspan="2">Basidiomycetes</th><th colspan="2">Other fungi</th><th colspan="2">Bacteria and actinomycetes</th></tr>
<tr><th>living</th><th>total</th><th>living</th><th>total</th><th>living</th><th>total</th></tr>
<tr><td>Woody debris</td><td></td><td>30.5</td><td>216.9</td><td>7.3</td><td>34.7</td><td>2.6</td><td>601.6</td></tr>
<tr><td>L</td><td rowspan="4">depth 36 cm</td><td>3.1</td><td>8.7</td><td>0.5</td><td>4.1</td><td rowspan="4">37.3</td><td rowspan="4">8433.3</td></tr>
<tr><td>(O_h + A_h)</td><td>8.9</td><td>31.7</td><td>3.4</td><td>12.9</td></tr>
<tr><td>A</td><td><1.0</td><td><1.0</td><td>26.4</td><td>97.5</td></tr>
<tr><td>B</td><td><1.0</td><td><1.0</td><td>31.4</td><td>155.6</td></tr>
<tr><td>Dead roots</td><td></td><td>228.0</td><td>1628.5</td><td>65.1</td><td>325.7</td><td>8.0</td><td>1851.1</td></tr>
<tr><td>Total</td><td></td><td>271.5</td><td>1886.8</td><td>134.1</td><td>630.5</td><td>47.9</td><td>10 886.0</td></tr>
</table>

In mull organic matter the same principle applies (Table 7.4). Again basidiomycetes have the highest living biomass in the decomposing litter horizons with lesser amounts where minerals are mixed with the finely divided organic matter in the A horizon, where other fungi are dominant. Notice also the very high live biomass of fungi, compared with bacteria, in the larger pieces of woody debris and roots where the ability to penetrate is important, whereas bacteria have their highest live biomass in the particulate matter. In mull humus, as noted earlier (p. 155), bacteria are more important than in mor, but their live biomass still does not exceed the live biomass of the fungi.

In agricultural systems, with their generally low levels of organic matter, basidiomycetes seem comparatively rare, or at least are not much isolated and they do not fruit on arable land or on ley grass systems that are periodically disturbed. Field mushrooms (*Agaricus campestris*) are characteristic of permanent pastures and meadows.

Protozoa

Protozoa are important inhabitants of the guts of various insects which live on plant remains. As free-living organisms they are found in all environments, though the type of vegetation and the rainfall influences the species present (p. 156) and their activity. Some can produce cellulase and so may be saprovores but most feed on bacteria: this is said to greatly increase the turnover of the bacterial population and the rate of mineralization. There is good experimental evidence, mostly from laboratory experiments, that this is so. However, it has been disputed by some researchers who point out that protozoa, and arthropods, eating bacteria should prolong immobilization for the length of time that the protozoan lives. It all depends on whether the bacteria would normally die and release nutrients, in which case the immobilization would be prolonged by being eaten, or whether the bacteria would exist for a long time in a dormant state if not eaten by the short-lived protozoan. The latter is more likely, hence grazing usually increases mineralization.

There are different communities associated with different litter types. Even in moss litter which often forms peat, there are many amoebae and various predatory species including suctoria. In grass litter and deciduous forest there is a more diverse community including ciliates and flagellates representing all the different feeding strategies from saprovores through bacterivores and fungivores to predators (Fig. 7.3). Litter in conifer forest, where the environmental factors are usually less favourable, does not have such a diverse or numerous fauna. Some of these characteristics are illustrated in Table 7.5 where different vegetation types should be compared within the same climatic region. Thus there are more ciliates in the deciduous woods than in conifers in Alaska, New Hampshire and Louisiana, and the highest proportion of ciliates with the highest pH. The grassland next to pine forest again shows an increase in ciliates and the grazed pasture a further decrease in the proportion of testacea.

Nematodes

Apart from being plant parasites many soil nematodes live on bacteria, fungi and protozoa. Their number generally relates to that of their prey and therefore varies in different soils or horizons. Biomass is usually greater (by an order of magnitude) in mull humus than under conifers (see Table 7.1). Again numbers peak in the amorphous humus layer or in the top of the A horizons where organic matter and minerals mix. Fungivorous and bacterivorous nematodes may have a significant effect on the biomass of their food organisms. Grassland and arable soils and litters have comparatively low nematode biomass and these are bacterivorous types. Forests, particularly deciduous ones, may have a slightly higher biomass and relatively more fungivores.

Microbes associated with saprotrophic invertebrates

Arthropods have a variety of cutting and chewing mouth parts to comminute the litter, snails and slugs grind it away with their rasp-like radula and earthworms can tear leaves apart. There may be further grinding in the gizzard or crop of some animals. The smallest members of this group, the collembola and mites, may feed directly on the higher plant, algal or bryophyte remains, though many eat the fungi and bacteria growing on the litter rather than digesting the litter itself. Certainly few can live on freshly fallen litter, it has to be partially decayed before it can be utilized. Most of these animals do not possess cellulase or lignin-degrading enzymes and so are dependent on the external microflora to degrade these portions of the primary resource. The fungi and bacteria represent a secondary resource much richer in nitrogen than the original leaf. There is however very great variation between different animal species in their food preferences and in some cases there is evidence of specialized feeding on particular microbial species, so affecting for example the makeup of fungal communities or some beetles feed only on myxomycete spores and so 'graze' these particular organisms.

Apart from these effects of grazing on the microflora of decaying litter the most important effect of arthropods is to break up the litter into small pieces,

Table 7.5 Protozoa in litter from different vegetation types in different climatic regions. Loss on ignition is approximately equal to organic matter loss. (Selected from Bamforth, S.S., 1971, *Journal of Protozoology*, **18**, 24–8.)

Location and climate	**Vegetation**	**pH**	**Moisture** (%)	**Loss on ignition** (%)	**Bacteria** (10^6 g^{-1})	**Protozoa number** (g^{-1})	
						Ciliates	Testacea
Alaska	Spruce, peat	6.1	60	48	12	100	2400
cold temperate	Birch, loam	5.8	56	65	25	500	9000
New Hampshire	Pine-hemlock, loam	4.2	47	87	29	1500	13200
cold temperate	Beech-maple, loam	7.6	59	85	24	5000	1000
Louisana	Pine, loamy-sand	4.3	32	93	73	500	18000
sub-tropical	Oak-hickory, silty-loam	4.7	54	88	91	3000	9000
	Pine, silty-loam	3.8	45	85	220	200	18000
	Grass-pine, next to pine above, silty-loam	4.8	47	60	370	1750	8000
	Grazed pasture, silty-clay-loam	6.5	24	62	410	1200	1000

usually down to only a few micrometres diameter in the case of small collembola. This increases the surface area available for microbial, especially bacterial, colonization. Small animals also disperse the spores of micro-organisms.

Those animals which apparently feed on the plant litter itself must have micro-organisms in their guts, usually symbiotic bacteria, to degrade the cell wall polymers. They have some part of the gut greatly expanded in order to increase the residence time for the digestion of cellulose and to provide the space for their symbiont population. Millipedes have an enlarged mid-gut and many are also coprophagous so they reprocess faecal material with its rich supply of microbial protein. Termites have an expanded hind-gut and rectal pouch which contain their symbionts and some mites have a very large caecum.

These specialized gut floras have been most studied in termites; those of the New World, the lower termites, have a hind-gut which can be two thirds of the animal's weight. The microbiology is very complex. Firstly, as in all guts, there are many free-living bacteria all of which are strict or facultative anaerobes (*Streptococcus*, *Bacteroides*, spirochaetes) and may hydrolyse cellulose. Some can also fix nitrogen which may be very important, since cellulose hydrolysis and the environment in general where the termites live, may be nitrogen limited. The nitrogen economy of termites is further complicated by the use of the uric acid excretory products by the hind-gut microflora, so giving an almost closed nitrogen cycle. Furthermore termites often eat dead members of the colony and also the cast exoskeletons rich in nitrogen-containing chitin.

The second major microbial population in the hind-gut of termites is protozoa which are responsible for most of the cellulose digestion. Flagellates such as *Trichonympha*, *Trichomitopsis* and *Mixotricha* occur in different termites. Protozoa, as eukaryotes, should not live in the anaerobic hind-gut: they do so by a remarkable collection of bacterial symbionts. There are symbiotic spirochaetes attached to special organelles on their surfaces which give motility by beating like cilia, ecto- and endosymbiotic bacteria, and sometimes intranuclear symbionts as well. The function of some of these is unknown but they may replace mitochondria in anaerobic habitats and may produce cellulase, and possibly lignin-degrading enzymes.

Numbers of bacteria and protozoan symbionts are very high: 10^8–10^{10} spirochaetes ml^{-1} and protists 10^7–10^8 ml^{-1} of hind-gut fluid. The hind-gut allows the exploitation of a difficult food source and in the case of New World termites makes them almost the only herbivores in some desert systems. The gut flora of other insect herbivores has not been extensively studied though cockroaches are known to harbour a large bacterial flora which degrades cellulose anaerobically to volatile fatty acids, and some beetle and crane-fly larvae may have significant gut symbionts.

An alternative strategy adapted by some New World ants (*Acromyrmex*) and Old World termites (Macrotermites) is to grow micro-organisms on specially collected 'litter' in fungal gardens (Batra, 1979). Instead of eating microbes on natural litter they cultivate pure cultures of one or a few species of fungi (*Lepiota*, *Xylaria*). This becomes their main food source, and may, under the influence of the ants or termites, produce specialized protein-rich hyphal structures which are eaten. The insects are improving the resource quality of the leaves by allowing the fungus to hydrolyse and respire the cellulose and sometimes the lignin, while immobilizing the nitrogen, hence lowering the C:N ratio. The problems involved in this association are formidable, for the pure cultures are grown in holes in the ground with a volume of up to 30 litres using natural leaves to make a compost. This shows a degree of sterile technique better than most microbiologists who employ specialized sterile media, antibiotics, disinfectants, autoclaves and sterile transfer cabinets! How do the insects do it? There is no doubt that without the ants or termites the cultures would rapidly become contaminated with 'weed' moulds and bacteria. Firstly the insects clean the collected pieces that they use; one species of ant generally uses one main sort of substrate, be it tree leaves, flower parts or even faecal pellets of caterpillars and sometimes the bodies of insects as well. Whatever the material it is carefully cleaned and then chewed to make a pulp. The ants themselves also have remarkably clean body surfaces, even though they get contaminated if they forage outside the nest. They have special structures to aid in cleaning themselves. This implies that they can remove (and sense the presence of) objects the size of bacteria and fungal spores. The pulp may have various secretions added, including saliva and anal droplets which contain nitrogen to improve the resource quality and in some cases proteinases to supplement the fungal cellulases and lignin-degrading enzymes. Maybe the secretions also contain antibiotics to control growth of foreign moulds. The 'compost' and 'gardens' are also physically weeded to remove contaminants. These operations again suggest that the ants must be able to detect and possibly even pick up, particles the size of bacteria and fungi.

The larvae and the adults feed on the fungus which is carried to new colonies by the emerging virgin queens in the swarms. The exhausted substrate is carried from the nest and dumped, often in special tips. It forms a rich fertilizer, adding minerals to the soil along with the organic matter. The ants may be very common in some situations and are of agricultural importance when they strip citrus trees of leaves.

The termites are a major pest of forest, orchards, agricultural crops and the timber in houses and furniture. The fungi are *Termitomyces* (a basidiomycete known only from termite mounds) and *Xylaria*. Pure cultures are grown, or a stable mixture of the two fungi, on a compost composed of the faecal pellets of the termites which have been eating cellulosic substrates during foraging. The material must again be sterilized in some way to enable the gardens to be kept pure.

There are many other invertebrate–fungal associations of varying degrees of sophistication. The ambrosia beetles which inoculate their brood galleries have already been mentioned in connection with timber decay (p. 96) and wood wasps also carry fungi and inoculate the tree at the time of egg laying. All these systems are concerned with the use of the fungus to digest cellulose for which the insects do not possess the appropriate enzyme systems.

Microbes associated with the saprotrophic vertebrates

Herbivorous vertebrates have the same problem as invertebrates: they cannot themselves digest the major component of their diet, cellulose. Again they use micro-organisms to do it for them in expanded regions of their gut, usually either an expanded oesophagus (rumen) or an enlarged caecum. Most of the studies have been done on agriculturally-important ruminants (sheep and cows) but it is thought that the process in wild animals is basically similar. There have been many reviews and books written on the rumen and its microflora and the biochemistry under different physiological conditions, and with different food resources, is now well understood (Hobson & Wallace, 1982).

The rumen develops as the young animal is weaned and starts to take in considerable quantities of plant material. It obtains bacteria and protozoa from the saliva of its mother or siblings during licking and from faecal contamination. The rumen has a large volume (about 100 l in a cow) and hence a long residence time for the contents which are returned to the mouth for further chewing from time to time. Liquid is supplied by drinking water and by the saliva (about 200 l day^{-1} in the cow) which is buffered and contains minerals, and urea as a nitrogen source. The bacteria and protozoa digest the cellulose and are themselves digested by the animal in the rest of the gut. Fatty acids, produced by the fermentation are absorbed by the animal and ammonia from deamination of amino acids and urea is used by the microbes. These reactions are summarized in Fig. 7.9. The only other product of importance is methane produced by *Methanobacterium ruminantium* from hydrogen which results from the anaerobic fermentation. This is a loss of carbon and energy from the system but does remove the hydrogen.

The bacteria are very numerous (10^9 ml^{-1}) and are all strict anaerobes such as *Streptococcus bovis*, *Ruminococcus flavifaciens*, *Bacteroides succinogenes*, *B. ruminicola* and *Selenomonas ruminantium*. No one bacterium is responsible for the rumen function, some breakdown starch (*B. amylophilus*) others convert pyruvate to propionic acid (*S. ruminantium*) and so on (see Hobson & Wallace, 1982). The precise composition of the flora and the biochemical reactions, are determined by the feed; animals fed on high protein concentrates have a different flora to those on hay. It is therefore important to change the diet slowly to allow the rumen microflora to adapt.

The protozoa are present at about 10^3–10^4 ml^{-1} and are mostly flagellates and ciliates. Typical species are *Isotricha prostoma*, *I. intestinalis* and *Entodinium caudatum* and they may have special organelles for attachment to fibres of plant material. *Entodinium* may ingest the end of fibres much larger than itself and digest them continuously while it is attached. Protozoa also digest proteins and many ingest fragments of plant material and starch grains. Despite all this however, ruminants can manage very well without their protozoa (though not without bacteria). Animals free from protozoa are apparently healthy and may grow as fast as, or even faster, than animals with the normal flora and fauna.

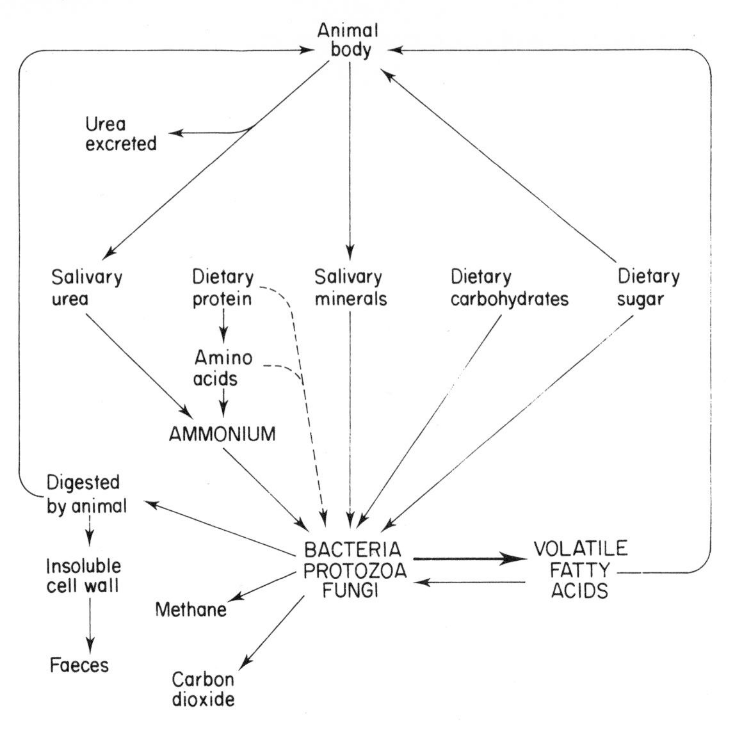

Fig. 7.9 The main reactions carried out by micro-organisms in the rumen. (From Campbell, 1983.)

There are also anaerobic fungi in the rumen and caecum which are apparently Mastigomycotina with flagellate zoospores (e.g. *Neocalimastix frontalis* and *Piromonas communis*). They are attached to, colonize and degrade plant material but their importance is at present unknown. Various other symbiotic associations have been suggested, but not thoroughly investigated. Methanogenic bacteria have been shown to be consistently present on the surfaces of some protozoa. They may be using carbon dioxide and/or hydrogen from the protozoan as their substrate. So far as is known there are none of the very complex symbiotic associations found in termite gut.

The organisms associated with plant litter decay are therefore very variable and range from free-living saprotrophs, through various loose associations with invertebrates and their faecal pellets to full mutualistic symbionts in the guts of invertebrates and vertebrates. It has long been recognized that 'detritus food chains' are based on microbial decomposition of litter, but it is also true that all grazing chains are dependent upon microbial breakdown of plants in the hind-gut, rumen or caecum. The crucial step in food chains from plant primary producers to any sort of animal is governed by the microbes which can degrade cellulose and lignin.

Horticultural and agricultural fermentations of plant litter

There are four main systems to be considered: firstly, horticultural and garden composts where the aim is to decay the material to a humus-rich soil additive. Secondly, there is the use of composting to destroy urban waste and refuse. Thirdly, the production of a special compost such as that used for growing mushrooms commercially. Finally, there is silage making, where the aim is entirely different: conditions are arranged so that the material is preserved with the minimum of decay and nutrient loss.

Horticultural and garden composts

It should be made clear that we are talking about the serious production of garden composts, and not about the garden rubbish heap. Generally there is a mixture of non-woody plant material, piled into a large enough heap to retain the heat produced by the very rapid decay, so that weed seeds and undesirable microbes are killed. Ideally the heap is turned to ensure that all parts are heated. Lime may be added to control pH which tends to fall during decomposition and frequently some form of available nitrogen is given to reduce the C:N ratio.

The initial colonization is by mesophilic fungi and bacteria (and probably protozoa though there is no information) such as deuteromycetes, ascomycetes and the common soil bacteria. As the temperature rises thermophilic fungi dominate (*Mucor pusillus*, *Chaetomium thermophilie*, *Humicola insolens*, *Aspergillus fumigatus* and *Thermomyces lanuginosus*) and these are finally replaced by thermophilic bacteria and actinomycetes above 55–60°C after 4–6 days. All organisms, including thermophilic fungi, may be killed in the hottest parts of the compost but they later re-invade from around the cooler edges of the heap. Thermophilic actinomycetes seem to have a very patchy distribution and their numbers vary greatly, though they do peak at the maximum temperature.

During this process there is a loss of cellulose, hemicellulose and other plant polysaccharides, while nitrogen tends to be conserved (Table 7.6) and soluble forms are immobilized: there is almost no free ammonium or nitrate after 35 days.

This description has not taken account of any animals and they are not thought to be important. They are found only around the cool edges or in heaps that have been incorrectly made so that they do not heat up. They may invade the heap after the initial rapid decomposition.

Composting urban waste

There are several reasons for composting urban waste: it reduces the bulk of the rubbish and produces a useful end product. Urban waste is becoming increasingly compostable as the amount of ash and dust drops and paper and card packaging waste increases. In 1975 domestic refuse in the U.K. was approximately 18% dust and ash, 50% paper, 13% vegetable waste and 16% glass, metal and plastic. Additional organic matter and nitrogen may be added

Table 7.6 Analysis of compost during its decay. (Modified from Yung Chang & Hudson, H.J., 1967, *Transactions of the British Mycological Society*, **50**, 667–77.)

	Days of composting				
	0	5	16	34	60
Fractions as % initial dry wt					
Total	100.00	86.68	73.48	48.47	48.79
Hemicellulose	35.69	32.76	26.55	16.89	16.98
Cellulose	45.32	35.04	30.30	11.15	13.27
Lignin	9.59	7.20	7.81	11.42	10.08
Diastase soluble	7.08	9.61	7.65	7.51	7.03
Ethanol soluble	2.30	2.16	1.17	1.50	1.41
Fractions in mg 100 g^{-1} dry wt					
Total N	827.6	789.7	532.6	628.9	651.8
Ammonium and nitrate N	838.1	450.5	12.7	5.2	4.3
Temperature °C	15.0	68.0	49.0	16.0	18.0

by using sludge remaining after the treatment of sewage; this improves the compost, mainly be increasing the nitrogen level, and also saves pollution of a river or estuary into which the sludge would otherwise be dumped.

The process is microbiologically similar to garden composts described above, and heaps or containers are used with frequent turning or mixing. After the initial rapid microbial decomposition lasting a few weeks the compost is matured for several months. The problem with urban waste is noncompostable impurities and, especially if sewage sludge is used, the presence of heavy metals which while not important in individual batches may build after continuous use over many years. The rubbish must first be passed through screens to remove large items, and then the ferrous metal is removed magnetically. Plastics and nonferrous metal are removed either by hand or by utilizing density differences with centrifugal separators. The remaining material is shredded, wetted, sludge added if this is used, and then the whole is thoroughly mixed. Composting then takes place. The end product may be screened again or shredded and nitrogen, phosphorus and potassium fertilizer may be added to improve the nutritional balance. The urban waste composts tend to have lower organic matter and minerals than garden composts and have a higher ash content. These processes are summarized in Fig. 7.10.

There is a future for composting urban wastes. Land-fill, a traditional way of disposal, is becoming increasingly difficult and expensive because of the lack of sites near large connurbations. Incineration requires very expensive equipment, can produce unacceptable fumes and is expensive to run because oil or gas usually has to be used to burn the wet rubbish at a high enough temperature.

Mushroom composts

Agaricus bisporus, the commercially cultivated mushroom in Western Europe, American and Australasia, is grown on straw-based compost and

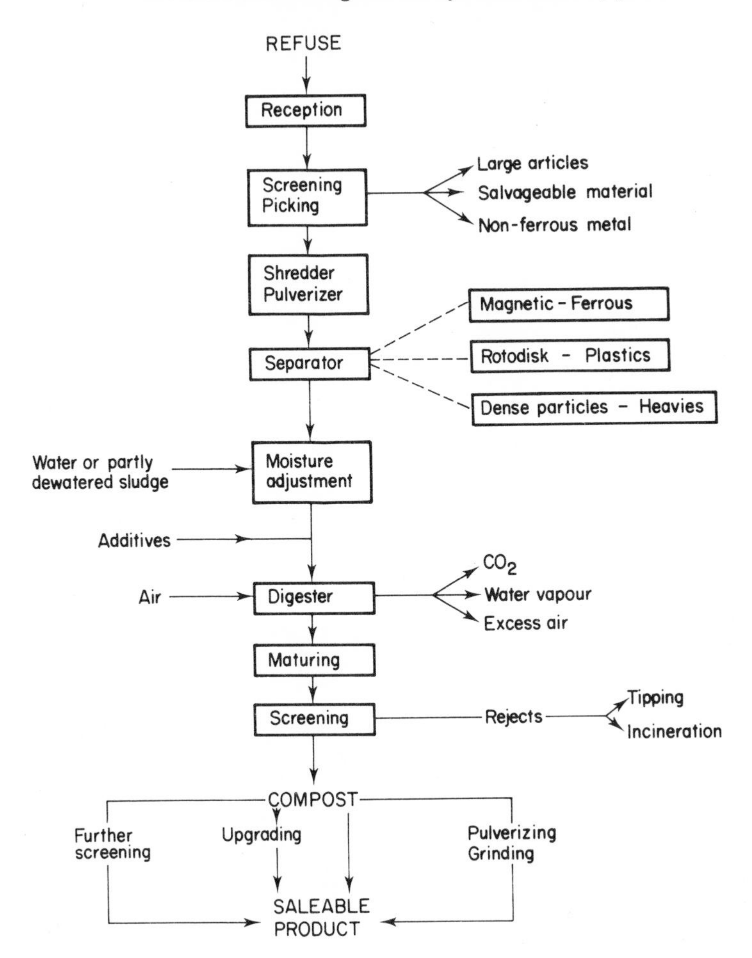

Fig. 7.10 Flow diagram of a plant for composting urban waste. (From Gray, K.R. & Biddlestone, A.J., in Dickinson, & Pugh, 1974.)

traditionally horse manure has been used. The straw is enriched with either inorganic nitrogen or with chicken manure, or with other high nitrogen organic material, and is then stacked in rows 2 or 3 m across the base and 2 m high. These rows can be turned and aerated by machines which straddle the row and move slowly along as they mix the compost (Fig. 7.11 A). There is a rapid growth of microbes, and if properly managed, the heating of the entire aerated stack. Additional aeration may be supplied by ducts within the stack. The microbiology is again dominated at this stage by thermophiles, especially actinomycetes. The high temperature should kill most other fungi although the compost may be chemically or steam sterilized in the trays in which the mushrooms are to be grown. Alternatively the trays of compost may be allowed to heat up naturally in hot rooms. Whatever method is used the compost, with a much modified microflora, is then inoculated with *A. bisporus* produced in

Fig. 7.11 **A.** Straw being prepared for mushrooms compost. The machine at the far end of the row moves along, turning the heap as it passes over it. **B.** Composite picture of a tray at different stages of production. (1) The prepared compost with inoculum of *Agaricus bisporus*. (2) Casing layer added. (3) The first flush of mushrooms just showing through. (Photographs courtesy of David A. Wood, Glasshouse Crops Research Institute, England.)

pure culture, usually on sterilized grain (Fig. 7.11 B). Growth of the mycelium takes 14 days at about 25° C and the compost is not sterile during this time. When the compost is full of mycelium, fruiting is induced by 'casing' it, that is covering it with several cm of soil or peat with chalk or limestone added (Fig. 7.11 B). After some days the mycelium grows through this low nutrient casing and the temperature is reduced to about 10 °C and fruit bodies start to form about one week later (Fig. 7.11 B). During casing the bacterial numbers in the soil rise, especially pseudomonads, and this is essential for the initiation of fruiting. Fruit bodies are produced in flushes for several weeks and when production falls off the compost is discarded.

This is the most complex form of composting plant materials, in which the formation of a good compost is only the start of the real process. Many of the methods of production are traditional but are now understood to some extent, though it is unlikely that any artificial system of growing *A. bisporus* will replace the methods of straw composting described above.

Silage

The whole object of the exercise and the microbiology is different in this process: special efforts are made to establish and keep anaerobic conditions, to prevent heating during the fermentation and to reduce the amount of plant material which is decomposed.

Silage may be made from many different plant materials: grass, maize and forage legumes are used and almost any plant material with a reasonable nutrient and dry matter content is a possible substrate for silage production. The presence of easily available sugars is necessary to make sure that the fermentation gets off to a rapid start before spoilage occurs. The plant material is harvested and should in the process be bruised and chopped so that it will form a compact heap. It may be put into specially built silos or into heaps packed down firmly and covered with polythene or into large polythene bags, to exclude air.

The main organisms are *Lactobacillus*, *Streptococcus* and lesser numbers of *Pediococcus* and *Leuconostoc*. Glucose is converted to lactic acid. Grass when alive has very low numbers of these lactic acid bacteria (less than 10 to 100 cells g^{-1}) but they soon become dominant. *Xanthomonas* and *Pseudomonas* may occur in the first three days or so and Enterbacteriaceae may exist for up to 7 days along with *Lactobacillus curvatus* and *Lactobacillus plantarum* but the lactic acid produced results in the acid-tolerant *Lactobacillus buchneri* and *Lactobacillus brevis* then becoming dominant. Fungi and other aerobic organisms are excluded, though *Clostridium* may occur if the acidity is not produced quickly enough. Clostridia use nitrogen and convert it to toxic amines and they may also produce butyric acid which spoils the silage. The grass, or whatever, may be wilted prior to ensilage since over-moist material encourages *Clostridium*. Various additives may also be used: molasses increases the sugar content to get a rapid initial fermentation. Mineral acids lower the pH and organic acids do the same, but also inhibit fungi but not lactobacilli. Formic acid drops the pH and also inhibits butyric acid-producing clostridia. It is also possible to buy commercially produced *Lactobacillus* of suitable strains or mixtures to establish a successful fermentation. Once anaerobic, acid conditions are produced and maintained the silage will keep for long periods without loss of nutritional value. However on opening the silo and admitting air, spoilage can be quite rapid with massive invasion by yeasts (e.g. *Mycoderma* and *Torula* at 10^{12} g^{-1}). Filamentous fungi may then follow including those such as *Aspergillus fumigatus* which can produce mycotoxins (p. 44).

This is again a case where the basic fermentation system is understood and there is sufficient practical knowledge of the necessary dry weight content and the effects of additives to produce good or at least acceptable silage in most years. There is however still much more information needed before the process is understood well enough to control it by other than the empirical and rather crude methods used at present.

Industrial fermentations of litter

There are many industrial fermentations dependent upon plant products ranging from beer and wine production, through organic acids and antibiotic production which are dependent upon molasses, sugar or starch as a carbon source, to very specific pharmaceutical products which may be produced from plant sterols. These will not be considered here since they would form a whole book in themselves (see Reed, 1982). We will restrict this discussion to the fermentation of crude plant material, often waste products of agriculture, which is recognizably 'plant litter' rather than seeds, fruits or refined plant products.

Though there is much theoretical information, and it should be possible to use lignin and cellulose-hydrolysing organisms on plant waste, there are rather few practical processes which use resources of such low quality (Smith, Berry & Kristiansen, 1983). Research is currently going on in three fields, the production of single cell protein and microbial biomass, the production of ethanol and the production of methane.

Protein production

Single cell protein production based on sugar-cane waste, cassava, straw and rice is usually performed with mixed cultures of fungi. The organisms cannot be separated from the substrate, and though the protein content may be quite high, it is not possible to produce a protein flour or powder by this method. Such products require fermentation on pure substrates. The crude fermentation does however upgrade the waste into a more protein-rich feed. *Trichoderma reesii*, *T. lignorum* and *Chaetomium cellulolyticum* have been used amongst other organisms. These methods will doubtless be developed, especially in the poorer countries, to provide badly needed animal feed (instead of using grain or legumes suitable for human use).

Liquid fuels from plant biomass

The production of ethanol from plant waste has been known to be possible for many years but the yields are quite low and recovery of the product is difficult. The alcohol can be produced from petroleum products or refined plant waste, such as molasses, more easily and more cheaply. The use of alcohol as a petroleum replacement has spurred further developments however, and costs are beginning to be competitive. Many of the systems depend on mineral acid hydrolysis of the cellulose and then fermentation of the low level of sugar produced (less than 15%). This limits ethanol production to no more than

7.5% (w/v) and there is then an expensive distillation process to recover this small amount. One possibility is the use of cellulases, produced for example during biomass production on another cellulose substrate, for the initial hydrolysis in the ethanol fermentation sequence. The direct conversion of cellulose to alcohol has been done and some likely organisms are being tested (e.g. *Clostridium thermocellum* and *Monilia* sp.). As oil and oil-based products increase in price these processes may become competitive, but it seems more likely that some plant sugar waste will need to be used as a substrate, rather than the bulky fibrous plant litter materials from which it is difficult to get sufficiently concentrated fermentable sugar.

These various possibilities are summarized in Fig. 7.12. Most suggested routes depend on non-microbial pre-treatment of the biomass and a mixed culture or sequence of different organisms is used. It is estimated that fuels from biomass may be potentially able to produce up to 10% of the U.S. energy needs and development of feed stocks, such as acrylic acid may become essential as petroleum ceases to be available at an economic cost.

Methane

The other approach to the rising costs of oil-based fuels is to produce methane by the anaerobic digestion of plant litter. The normal substrate for this is liquid waste high in organic matter, usually animal or human sewage slurries, where there is adequate nitrogen and the bacteria produce methane by reducing carbon dioxide or organic carbon. However the production of methane from plant litter is possible as is obvious from the production of 'marsh gas' in swamps and stagnant pools. It again requires a two-stage process, first to break down the polysaccharides and then to use the resulting fatty acid, acetic acid and carbon dioxide as a substrate for methanogenesis. Plant litters may be added to liquid slurry, but on their own they are too low in nitrogen and slow to decompose. A nitrogen source must be added and this may be ammonium sulphate or some other nitrogen-rich material. The substrates vary enormously and almost everything has been tried from forest residues, food processing wastes from canning factories, to sugar-cane waste (bagasse) on which commercial fermenters have been running for 20 or 30 years. Even water hyacinths, an introduced weed in many countries, have been used. This gives an incentive to clear a noxious weed and results in a useful product at the end. A further refinement of this is to use water hyacinth in sewage treatment and then to produce methane from the now nitrogen-rich plants. Another integrated system uses agricultural waste to ferment to alcohol which is used as fuel, the residue from this fermentation is anaerobically fermented to methane which is again used as fuel, and the final residue is used as fertilizer.

It seems that methane production is not likely to be a large industrial process on its own, using its own special crop. Its success has been in fermenters attached to particular factories producing suitable waste, or attached to small villages using mixed agricultural waste and sewage. The gas produced is then used on site either directly as a flame or by running a generator for electricity. The technology is therefore relatively simple and the fermenter can be constructed and used by unskilled labour in third world countries. The

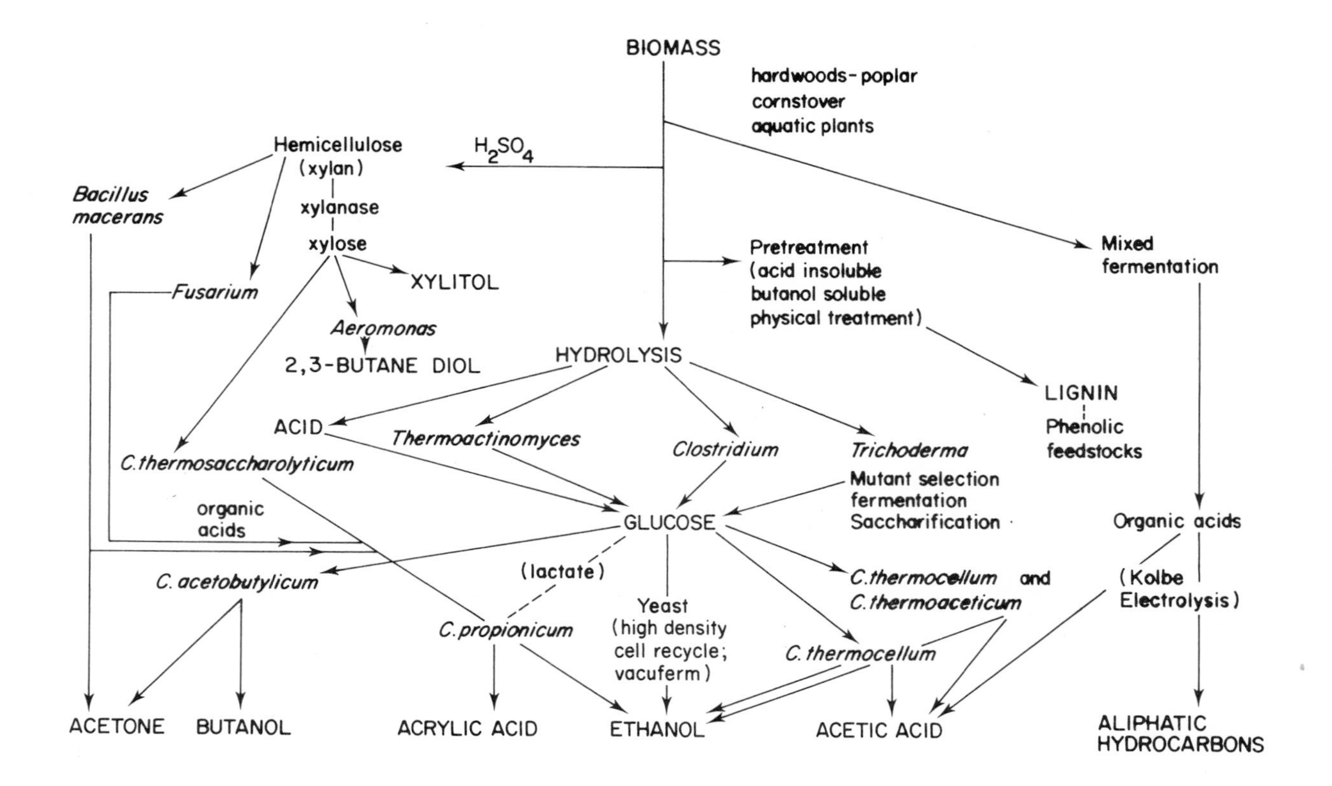

Fig. 7.12 Possible use of plant biomass to produce liquid fuels and industrially useful chemicals as envisaged by the Fuels from Biomass Program of U.S. Department of Energy. (From Jeffries, T., Montenecourt, B.F. & Eveleigh, B.E., in Smith, J.E., Berry, D.R. & Kristiansen, B., 1980, *Fungal biotechnology*. Academic Press, London. p. 225–42.)

potential is enormous: Table 7.7 shows an example from Latin America with considerable energy production and saving in wood consumption. The latter is important because it saves felling forest and creating erosion problems. Many hundreds of thousands or even millions of small, simple biogas fermenters using plant waste as part of their feed stock are in use worldwide, for example in India, Africa, China, Latin America, Korea and South East Asia.

The decomposition of plant litter is a difficult process because of the recalcitrance of the major polymers and the great mixture of substances present. Micro-organisms play a key role in all systems, both natural, agricultural and industrial because they are the only organisms producing cellulases and lignin degrading enzymes. As such they link the primary producers of material, the plants, with all other organisms. Microbes are also the key organisms in industrial handling of plant biomass and waste materials to produce useful products, to prevent pollution and to reduce man's dependence on non-renewable energy and industrial feedstock resources. Plant litter is the most widely available organic resource and is renewable in healthy ecosystems. Micro-organisms have been degrading plant litter for many millions of years and have only comparatively recently been used in fermentations and degradations for man's industrial uses; it is to be hoped that organisms can be found in the natural environment, or modified by man, to use unusual quantities or types of plant material and to produce novel compounds.

There are hopes, probably valid, that the remains of unusual plants, such as latex-rich *Euphorbia*, may be able to replace some of the present dependence on oil. Alternatively common plant waste may, by unusual combinations of microbes or developments in fermentation technology, be made to yield products now produced from oil (see Fig. 7.12). Care is however needed. At present the carbon cycle is being upset by the sudden use of fossil fuels and the generation of unnatural quantities or types of chemicals; there is widespread contamination of all habitats by many types of pollutants. The best feature of using plant biomass and waste litter is that it is 'natural' and 'local'. The temptation must be avoided to make it the subject of high technology, to make biomass consumption so large that special crops are needed or forests are felled to supply large industrialized fermenters. This would create more problems than were solved, by overexploiting even plant waste and litter which in the long term is needed by soils. The complex activities involved in plant litter incorporation in soil are essential for its productivity.

There is the opportunity, as the end of the oil-based economy is in sight, to try to replace it with a system using renewable resources based on relatively small, integrated industrial facilities, and communities that use their own waste to produce some of their energy requirements. Such systems might also generate usable microbial biomass for animal feed to save more valuable plant commodities (grain, protein-rich legumes) for direct human consumption. It is not possible to replace our dependence on oil immediately or solve world food problems overnight, but a better utilization of the large amounts of plant biomass which is currently considered waste might at least make a start on these problems. So research into these various fermentations is worth pursuing since the basic technology and microbiology are already used in many traditional small scale production systems.

Table 7.7 Potential of biogas as fuel in rural areas of some Latin American countries. (From DaSilva, E.J., Shearer, W. & Chatel, B., 1980, *Impact of Science on Society*, **30**, 225–33. © Unesco 1980. Reproduced by permission of Unesco.)

Country	**Area** (km^2)	**Population, 1976** (percentages are those concentrated in rural areas)	**GNP from agricultural sector 1975** (%)	**Agricultural wastes** (thousand tonnes)[1]	**Equivalent production of biogas** (m^3 × 10^6)		**Energetic value** (kJ × 10^9)	**Deployment of biogas**[2] **in rural sectors** (%)	**Substitute for firewood per annum** (million kg)
					Total	Usable			
El Salvador	21 393	4 120 000 (59.0%)	24.2	912	104 458	74 521	1 558 971	16.8	456.30
Guyana	214 970	810 000 (1977) (58.5%)	34.2	273[3]	40 950	28 666	599 689	34.0	518.30
Honduras	112 082	3 040 000 (66.6%)	29.0	468[4]	53 279	37 295	780 224	28.22	625.25
Uruguay	186 926	3 139 000	13.2	1,172[5]	179 020	125 300	2 604 749	313.57	Complete

1. From rice, maize, sorghum, beans, green coffee.
2. Total from agricultural and livestock wastes.
3. From rice, maize, green coffee.
4. Inclusive of wheat residues.
5. From wheat, rice, barley, maize, oatmeal, sorghum, kidney beans, plant beans, seasonings, soybean.

Source: Organización Latinoamericana de Energia, Document PS/W144, 20 October 1979.

Selected references and further reading

Anderson, J.M. & MacFadyan, A. (Eds) (1976). *The role of terrestrial and aquatic organisms in decomposition processes*. Blackwell Scientific Publications, Oxford. pp. 474.

Batra, L.R. (ed.) (1979). *Insect-fungus symbiosis: nutrition, mutualism and commensalism*. Allanheld, Osmun & Co., Montclair. pp. 276.

Breznak, J.A. (1982). Intestinal microbiota of termites and other xylophagous insects. *Annual Review of Microbiology*, **36**, 323–43.

Campbell, R. (1983), *Microbial ecology*, 2nd edition. Blackwell Scientific Publications, Oxford. pp. 191.

Dickinson, C.H. & Pugh, G.J.H. (Eds) (1974). *Decomposition of plant litter*. Academic Press, London. Vol. 1. pp. 241 + 46, and Vol. 2. pp. 775 + 75.

Fenchel, P. & Blackburn, T.H. (1979). *Bacteria and mineral cycling*. Academic Press, London. pp. 225.

Hobson, P.N. & Wallace, R.J. (1982). Microbial ecology and activites in the rumen. *CRC Critical Review of Microbiology*, **9**, 165–225 and 253–320.

Persson, T., Baath, E., Clarholm, M. Lundkvist, H., Soderstrom, B.E. & Sohlenius, B. (1980). Trophic structure, biomass dynamics and carbon metabolism of soil organisms in a Scots pine forest. In Persson, T. (Ed.) *Structure and function of northern coniferous forests*. Ecological Bulletin (Stockholm), **32**, 419–59.

Reed, G. (1982). *Prescott & Dunn's industrial microbiology*, 4th edition, A.V.I. Publishing Company, Westport, Connecticut. pp. 883.

Smith, J.E., Berry, D.R. & Kristiansen, B. (1983). *The filamentous fungi*, Vol. 4. *Fungal technology*. Edward Arnold, London. pp. 401.

Swift, M.J., Heal, O.W. & Anderson, J.M. (1979). *Decompositon in terrestrial ecosystems*. Blackwell Scientific Publications, Oxford, pp. 372.

8
Future Prospects for Plant Microbiology

Many problems and possible areas of expansion in plant microbiology have been noted in the preceeding chapters. The aim is now to bring some of these points together, to try to develop some generalizations and to speculate on likely developments.

Present problems

The general problem in plant microbiology is a lack of basic knowledge; the seriousness of the deficiency varies from field to field but the urgency of some situations make it necessary to attempt to go straight into applied problems without the basic research. High on the list of difficulties is methodology, with no techniques giving complete answers to questions of enumeration of microbes in natural environments.

It is often very difficult to identify and classify microbes associated with plants, with the exception of a few well-studied pathogens and soil organisms. This makes the descriptive work either very tedious to do, or rather vague. The problem is perhaps worst with bacteria where it is often impossible to consider the task of naming isolates in any reasonably-sized study. Numerical taxonomy, factor analysis and principal components analysis are possibly solutions for the study of particular habitats, but make comparisons almost impossible. There is no way of telling whether the same organisms are being considered in different studies. In the long term, more theoretical study may produce a more easily used system, but this seems unlikely; if anything the identification is now coming to depend on more and more complicated tests and analyses with expensive equipment. In the near future plant microbiologists will just have to live with the problem.

Of the different groups of organisms the protozoa are the most neglected in the study of plant microbiology. This is partly because of particular methodological problems with protozoa, but mainly that there are just too few protozoologists. Too few studies have been done on protozoa to have any hope of arriving at a coherent picture.

Most work on plant microbes has been in agriculture, as was anticipated in the preface. This is quite reasonable because of the incentive of improved food production. It does however mean that many vegetation types are almost entirely negelected. Forests and desert vegetation perhaps suffer most and the former also poses technical problems with the size of trees giving extra sampling difficulties. Forests cover a considerable area of the land, in their many different forms, and some are highly productive. What little information we have suggests that microbes can be important in the primary production and nutrient flow in these ecosystems.

The problem is compounded in the tropics which in general are very poorly studied. There are hints of a very diverse flora, different from that of temperate countries. Communities on leaves for example could be having a great effect on productivity and apparently have a considerable biomass. If anyone feels in need of a challenge, of a new field to study, they should become a microbiologist, especially a protozoologist, specializing in tropical vegetation (if anyone can be found to train such investigators, and the habitat still exists by that time, there should be some interesting research papers!). There is a great need for baseline studies on the major vegetation types before they are irreversibly altered by man, either by direct exploitation or by pollution.

Fields of expansion in plant microbiology

It is only in the tropics that any really novel forms of plant microbes are likely to be found; perhaps there are still unknown, but important, symbiotic associations or organisms with an unusual complement of enzymes. Systems capable of simultaneously fixing nitrogen in significant quantities while degrading plant litter would be especially interesting in view of the availability of nitrogen-limited carbon substrates. There are for example reports of such systems from tropical wood-boring marine worms which apparently have unusual gut floras.

There is no clear solution to these gaps in our knowledge except a lot of hard work and a lot of investment of money and resources. This is not however likely to happen. The tropical third world countries, above all, are short of money and trained personnel: what they have of both of these scarce resources will be used, rightly, to combat the problem of producing food for the increasing numbers of their people. This is again short term research to solve immediate problems without the investment in basic knowledge for the future. Aid programmes from the so-called developed world are pathetically inadequate for even solving pressing problems, never mind seemingly obscure, pure research. It is not feasable to transfer sophisticated fermentation technology or microbiological processes requiring constant skilled monitoring. Most of the research in the developed world is improving the productivity in western agriculture which is already producing enormous surpluses which it will not, or cannot, transfer to the third world. How can people be persuaded to work in tropical countries, probably for less money, when they find more profitable and congenial work in their home countries? There will of course continue to be centres of excellence in the developing countries, but the major changes for the majority of the world's population will be the widespread use of simple technology based on microbiology (biogas generators are perhaps the best example of this) with some input from international agencies in providng new crop cultivars, etc.

Other growth areas will be in western agriculture, especially of commercially important crops. There are now experiments, in a very crude way, with all sorts of microbes which may control disease, produce growth-stimulating effects, improve symbiotic associations and prevent ice damage by applying bacteria which do not act as ice nucleation sites. There is often little basic understanding of how these organisms work, how best to apply them to crops and how to encourage them in natural and agricultural systems. A lot can be done very simply by the selection of wild organisms with better properties and there is also the whole Pandora's box of genetic engineering to produce organisms to suit particular habitats or to perform novel tasks. It is however extremely doubtful if we know enough about any plant-associated microbial community to actually design or improve an organism to fit into it. Basic research is necessary, and is now being done for nitrogen fixation for example, but for most applications in the near future it is likely that isolation and empirical strain selection still have much to commend them.

Microbiology is unlikely to be the cure-all for the world's food problems and too hopeful a picture should not be painted. The major cause is the expanding population and possibly the climatic changes which seem to be occurring. Microbiology (or any other technology such as plant breeding) can do nothing more than help to keep pace with minimal food demands. No doubt much more can be done by controlling microbial damage to crops and stored food, than by any 'magic bug' applied to the plant which is growing. It is estimated that about 30% of crops are destroyed during storage.

This brings us to consider plant pathology in its widest sense. Despite pathogenic or parasitic microbes, most plants remain healthy and most plant foods are not spoiled by micro-organisms, though they may be damaged also by animal pests. There is a real possibility of two strategies developing for plant disease control. One would be the control by chemicals, often by rather crude toxins, as at present. This may be cheap, or all that can be afforded, but there may be serious effects on non-target microbes or on the environment in general. The continued use of DDT in some third world countries is an example of this: even though its use is banned in countries that can afford the luxury of not using it, these same countries may still be manufacturing it. However DDT is undeniably a cheap and still very effective insecticide in some situations, and it may only have a short persistence in the tropics.

The path followed by the richer countries is likely to be different. Sophisticated chemical control methods are being developed to prevent microbial damage to plants: both the chemicals themselves and the method of application may require money and considerable technological resources and expertise. In the forseeable future microbes themselves will also be used in plant disease control on a serious scale, either by themselves or in integrated control systems. This all involves an understanding of the relationship between the plant, the microbial pathogen and the saprotrophs, and also a knowledge of the environment as a whole. An even more complex possibility is the control of microbial interactions by plant breeding, or possibly genetic engineering. It is already known that some gene changes carried out to produce resistant cultivars of wheat alter the rhizosphere microflora, though whether this is the cause of resistance or a fortuitous side effect we do not know. We do not understand anything about rhizosphere interactions in enough detail to design a change in the plant genome to cause a rhizospere effect to promote microbial control of some disease. Advances in biological and integrated control are likely to be much more pragmatic and empirical for a long while to come.

Microbial plant diseases are also beginning to be used to avoid or to supplement herbicides. Diseases can be used to reduce populations of weeds, and if an obligate biotroph is chosen it may well be species or even cultivar specific (e.g some rusts) so reducing effects on non-target organisms. Again it is not realistic to suppose that micro-organisms are going to take over weed control, but they may be useful. Herbicides will continue to be a major factor in increasing yield and productivity in those countries that can afford to use them.

Even where pesticides and fertilizers can be used on economic grounds there are some who say that they should not be. There is support for organic farming, using composts rather than fertilizer and avoiding pesticides as much as possible. This has already been discussed in connection with organic matter and microbial decomposition. Firstly it is clear that western agriculture with its high productivity based on fertilizers and agrochemicals could not sustain these yields with organic manures and no pesticides. If the developed world wishes to continue to eat to excess and to live in a manner to which it has become accustomed then the present agricultural system must be maintained. The third world cannot in general afford these methods, and it practices subsistence farming or uses organic manures in traditional systems.

A crop's requirements for nutrients can be met by chemical fertilizers: there is enough knowledge of both the major nutrients and trace elements to keep a plant healthy and growing. There is no evidence that microbial components of manures have any 'magic' effects and there is doubt about the efficiency of some specially designed microbial systems such as *Azotobacter* inoculation (p. 119). The main effect of organic manures is on soil properties such as crumb structure and no normal addition produces any dramatic changes (p. 161). However in the long term it is clear that organic matter in the soil must be eventually depleted if crops are removed and nothing is added back. This will then have adverse effects on the plants and micro-organisms because the soil's water holding capacity, etc., will be affected. Some crops grown in monoculture do not leave enough organic matter in debris and roots to replace that mineralized each year.

There could be a change in the situation if the costs of oil and therefore fertilizers, continue to rise. Alternative systems could then become more attractive and perhaps a thorough study of composting techniques would then be considered. It may be possible to use straw and other agricultural waste to compost and provide some nutrients for nitrogen fixation, so upgrading the value of the straw when put back on the land. The expensive and inconvenient handling of plant litter might then be worthwhile. Some of these systems were discussed in the last chapter for producing ethanol, methane and a final residue for manure and these could be attractive if energy costs continue to rise.

Organic farming cannot compete in yield production with modern farming techniques. It has a major role in those countries where a low capital and recurrent cost input is needed, even if yields are not always maximum. There may well be a future in organic farming even in the 'developed' western agriculture for reasons of soil structure or when the techniques of composting are improved or become competitive with rising prices. There would probably be lower yields, but most of western agriculture produces surpluses anyway.

The discussion of the extensive use of pesticides is more complex. As indicated above, biological control may become more important. There may be increased pressure against some pesticides for environmental reasons. Resistance to fungicides is a serious problem especially with the modern highly specific ones. There is also a finite limit to the number of suitable chemicals; perhaps many of the best ones have already been found and it will be more difficult in the future to find acceptable new formulations as resistance develops to the old ones. In the long term we may need, or have to have, integrated chemical and biological control.

The final thought should be that enough is now known to see many of the problems and gaps in our knowledge. We have the means to fill at least some of them. We must make better use of existing microbial technology and plant disease control measures rather than necessarily looking for high technology answers. Plant microbiology and the associated field of biotechnology is a growth area of enormous potential some of which is beginning to be realized.

Index